Study Guide to Accompany

Kolman ■ Denlinger

CALCULUS

Third Edition

For The Management, Life, and Social Sciences

Susan L. Friedman

Bernard M. Baruch College
City University of New York

Robert L. Higgins

Quantics, Inc. and Drexel University

SAUNDERS COLEGE PUBLISHING
A Harcourt Brace Jovanovich College Publishers
Fort Worth ■ Philadelphia ■ San Diego ■ New York ■ Orlando ■ Austin
San Antonio ■ Toronto ■ Montreal ■ London ■ Sydney ■ Tokyo

Copyright© 1992 by Saunders College Publishing

All rights reserved. No part of this publication may be reproduced
or transmitted in any form or by any means, electronic or mechanical,
including photocopy, recording or any information storage and retrieval
system, without permission in writing from the publisher.

Requests for permission to make copies of any part of the work should
be mailed to: Permissions Department, Harcourt Brace Jovanovich, Publishers,
8th Floor, Orlando, Florida 32887.

Printed in the United States of America.

Kolman/Denlinger: Study Guide to accompany CALCULUS FOR THE MANAGEMENT, LIFE, AND SOCIAL
SCIENCES, 3/E

ISBN 0-03-092733-1

234 021 987654321

0 Review of Algebra

Exercise Set 0.1 (Page 2)

1. For $x = 3$ and $y = -4$, $2x - y = 2(3) - (-4) = 10$.

3. $\dfrac{3}{8} \div \left(\dfrac{7}{12} + \dfrac{2}{3}\right) = \dfrac{3}{8} \div \left(\dfrac{7}{12} + \dfrac{8}{12}\right) = \dfrac{3}{8} \div \dfrac{15}{12} = \dfrac{3}{8} \times \dfrac{12}{15} = \dfrac{3}{10}$

5. 5 divided by 0 and $\dfrac{0}{0}$ are undefined.

7. $5x + 4 = 3x$ (Subtracting 3x from both sides
   $2x = -4$ and adding -4 to both sides)
   $x = -2$ (Dividing both sides by 2)

9. $x + 2y = 8$
   $3x - y = 17$
   $x + 2y = 8$
   $6x - 2y = 34$ (Multiplying second equation by 2 and adding)
   $7x = 42$
   $x = 6$ (Dividing both sides by 7)
   $6 + 2y = 8$ (Substituting x = 6 into the first equation)
   $2y = 2$ (Subtracting 6 from both sides)
   $y = 1$ (Dividing both sides by 2)

11. $16^{\frac{3}{2}} = \left(\sqrt{16}\right)^3 = 4^3 = 64$

13. $\left(\dfrac{a^{-2}b^3}{c^{-1}}\right)^3 = \dfrac{a^{-6}b^9}{c^{-3}} = \dfrac{c^3 b^9}{a^6}$

15.

x	-1	2	5	8
y	-4	-2	0	2

17. $2x^3 - 7x^2 - 15x =$
    $x(2x^2 - 7x - 15) =$ (factoring out x)
    $x(2x + 3)(x - 5)$ (factoring the trinomial)

2 Study Guide

19. The slope of the line is

$$m = \frac{y_2 - y_1}{x_2 - x_1} = \frac{2-(-1)}{4-2} = \frac{3}{2}.$$

Using the point-slope form with the point (2,-1) yields

$$y - y_1 = m(x - x_1)$$
$$y - (-1) = \frac{3}{2}(x - 2)$$
$$y + 1 = \frac{3}{2}x - 3$$

or

$$y = \frac{3}{2}x - 4$$

21. (a) line falls since its slope is negative.
    (b) y decreases as x increases since the slope is negative.
    (c) If x increases by 3 units, y increases by 3(-2) = -6 units—i.e. y decreases by 6 units.
    (d)

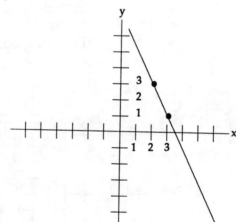

23. $2x^2 - 3x - 2 = 0$
    $a = 2 \quad b = -3 \quad c = -2$

$$x = \frac{-b \pm \sqrt{b^2 - 4ac}}{2a} = \frac{3 \pm \sqrt{(-3)^2 - 4(2)(-2)}}{2(2)}$$

$$x = \frac{3 \pm \sqrt{9 + 16}}{4} = \frac{3 \pm \sqrt{25}}{4} = \frac{3 \pm 5}{4}$$

$$x = \frac{3+5}{4} = \frac{8}{4} = 2 \qquad x = \frac{3-5}{4} = \frac{-2}{4} = -\frac{1}{2}.$$

Exercise Set 0.2 (Page 11)

3. $(4xy^2 + 2xy + 2x + 3) - (-2xy^2 + xy - y + 2) =$

   $4xy^2 + 2xy + 2x + 3 + 3xy^2 - xy + y - 2$

   Collecting like terms, we have

   $(4xy^2 + 3xy^2) + (2xy - xy) + 2x + y + (3 - 2) =$

   $7xy^2 + xy + 2x + y + 1$

7. $(2x + 3)(3x - 5) = (2x)(3x) + 2x(-5) + 3(3x) + 3(-5)$

   $= 6x^2 - 10x + 9x - 15$

   $= 6x^2 - x - 15$

15. $(2x - y)(x^2 + 2xy - y) = 2x(x^2) + 2x(2xy) + 2x(-y)$
    $\quad - y(x^2) - y(2xy) - y(-y)$

    $= 2x^3 + 4x^2y - 2xy - x^2y - 2xy^2 + y^2$

    Collecting like terms we have

    $2x^3 + 3x^2y - 2xy - 2xy^2 + y^2$

21. $25y^2 - 9x^2 = 5^2y^2 - 3^2x^2 = (5y)^2 - (3x)^2$
    $= (5y + 3x)(5y - 3x)$

23. $3x(2x + 5y) - 4y(2x + 5y) = (2x + 5y)(3x - 4y)$

    since $(2x + 5y)$ is the common factor of both terms.

29. $2x^3 - x^2 + 3x = x(2x^2 - x + 3)$

41. $$\frac{2a+3}{a^2+a-6}-\frac{a+2}{a-2}=\frac{2a+3}{(a-2)(a+3)}-\frac{a+2}{(a-2)}$$
$$=\frac{2a+3}{(a-2)(a+3)}-\frac{a+2}{(a-2)}\cdot\frac{(a+3)}{(a+3)}$$
$$=\frac{2a+3-(a^2+3a+2a+6)}{(a-2)(a+3)}=\frac{2a+3-a^2-5a-6)}{(a-2)(a+3)}$$
$$=\frac{-a^2-3a-3}{(a-2)(a+3)}=\frac{-(a^2+3a+3)}{(a-2)(a+3)}$$

45. $$\frac{a^2+a-2}{a^2+2a-15}\div\frac{a^2+2a-3}{a-3}=\frac{(a-1)(a+2)}{(a+5)(a-3)}\div\frac{(a-1)(a+3)}{a-3}$$
$$=\frac{(a-1)(a+2)}{(a+5)(a-3)}\cdot\frac{a-3}{(a-1)(a+3)}=\frac{a+2}{(a+5)(a+3)}$$

53. $$\frac{5-\frac{1}{x+4}}{\frac{3}{x+4}}=\frac{\left(5-\frac{1}{x+4}\right)(x+4)}{\left(\frac{3}{x+4}\right)(x+4)}=\frac{5(x+4)-1}{3}$$
$$=\frac{5x+20-1}{3}=\frac{5x+19}{3}$$

55. $$\frac{2}{x-1}+\frac{1}{3}=\frac{1}{x-1}$$

    We multiply both sides of the equation by the lowest common denominator 3(x - 1), obtaining

    $$3(2) + 1(x - 1) = 1(3)$$
    $$6 + x - 1 = 3$$
    $$5 + x = 3$$
    $$x = -2.$$

Exercise Set 0.3 (Page 23)

5. To find the slope of a line, we put the line into slope-intercept form y = mx + b. The coefficient of x will be the slope.

(a) $y = 3x + 2$. This line is in slope-intercept form and $m = 3$.

(b) $y = 3$. Since $y = 0 \cdot x + 3$, the slope $m = 0$.

(c) $x = \frac{2}{3}y + 2$

$x - 2 = \frac{2}{3}y$ or $y = \frac{3}{2}x - 3$

The slope $m = \frac{3}{2}$

7. When m is positive, the line rises from left to right; when m is negative, the line falls from left to right.

(a) $y = 2x+3$ $m = 2 > 0$ line rises from left to right

(b) $y = \frac{-3}{2}x+5$ $m = \frac{-3}{2} < 0$ line falls from left to right

(c) $y = \frac{4}{3}x - 3$ $m = \frac{4}{3} > 0$ line rises from left to right

(d) $y = \frac{-2}{5}x - 3$ $m = \frac{-2}{5} < 0$ line falls from left to right

15. We want to find the point-slope form for the line. This form is $y - y_1 = m(x - x_1)$.

(a) The given points are (-1,2) and (3,5)

The slope is $m = \frac{(y_2 - y_1)}{(x_2 - x_1)} = \frac{5 - 2}{3 - (-1)} = \frac{3}{4}$

We use this slope and either point in the form. Taking the point (-1,2), we have $y - 2 = \frac{3}{4}(x + 1)$

(b) The given points are (-3,-4) and (0,0)

The slope is $m = \frac{(y_2 - y_1)}{(x_2 - x_1)} = \frac{-4 - 0}{-3 - 0} = \frac{4}{3}$

Using the point (0,0) in the point-slope form yields

$$y - 0 = \frac{4}{3}(x - 0) \quad \text{or} \quad y = \frac{4}{3}x$$

17. We first solve $x + 2y = 3$ for y obtaining $2y = -x + 3$, or equivalently, $y = -x/2 + 3/2$. Hence, the slope of any line parallel to the given line is $-1/2$. Since the desired line passes through $(1,-2)$, we can use the point-slope equation with $m = -1/2$, $x_1 = 1$, and $y_1 = -2$ so that

$$y + 2 = \frac{-1}{2}(x - 1)$$

Multiplying both sides by -2 yields

$$-2y - 4 = x - 1$$

or

$$0 = 2y + x + 3$$

23. (a) $x + y = 6$
    $2x + 3y = 15$

    Multiplying the top equation by -3 and adding to the bottom equation gives $-x = -3$ or $x = 3$. Substituting $x = 3$ into the top equation, we have $y = 3$. Thus the only solution of the system is $x = 3$, $y = 3$, and the lines intersect at the point $(3,3)$.

    (b) $x - 2y = 7$
    $3x - 6y = 14$

    The system has no solution. There is no pair of real numbers x and y such that $x - 2y = 7$, then $3x - 6y = 14$, for if $x - 2y = 7$, then $3x - 6y = 3(x - 2y) = 3(7) = 21$. In this case the lines are parallel.

    (c) $x + 3y = 1$
    $-2x - 6y = -2$

    Since the second equation is -2 times the first, there are infinitely many solutions of the system and the lines are identical.

    (d) $x - 3y = -5$
    $-2x + 3y = -1$

0: Review of Algebra   7

Adding the two equations gives 3x = -6 or x = -2. Substituting x = -2 into the top equation gives -3y = -3 or y = 1. Thus the only solution of the system is x = -2, y = 1, and the lines intersect at the point (-2, 1).

Exercise Set 0.4 (Page 31)

5. $x^2 - 3x = 0$

   Factoring, we have

   $x(x-3) = 0$.

   Thus, either x = 0 or x - 3 = 0 so the solutions of the given equation are

   x = 0 and x = 3.

7. $2x^2 + 2x - 5 = 0$

   Using the quadratic formula with a = 2, b = 2, and c = -5, we have

   $$x = \frac{-b \pm \sqrt{b^2 - 4ac}}{2a} = \frac{-2 \pm \sqrt{(2)^2 - 4(2)(-5)}}{2(2)}$$

   $$x = \frac{-2 \pm \sqrt{4 + 40}}{4} = \frac{-2 \pm \sqrt{44}}{4} = \frac{2 \pm 2\sqrt{11}}{4}$$

   $$x = \frac{2(-1 \pm \sqrt{11})}{4} = \frac{-1 \pm \sqrt{11}}{2}.$$

13. $y^2 + 2y + 4 = 0$

    Using the quadratic formula with a = 1, b = 2, and c = 4, we have

    $$x = \frac{-b \pm \sqrt{b^2 - 4ac}}{2a} = \frac{-2 \pm \sqrt{(2)^2 - 4(1)(4)}}{2(1)}$$

    $$x = \frac{-2 \pm \sqrt{4 - 16}}{2} = \frac{-2 \pm \sqrt{-12}}{2}.$$

    There are no real solutions, because there is no real number whose square is -12.

21. $x^3 + 2x^2 - 8x = 0$
    $x(x^2 + 2x - 8) = 0$
    $x(x - 2)(x + 4) = 0$

Thus either $x = 0$, $x - 2 = 0$, or $x + 4 = 0$. Therefore, the solutions of the given equation are $x = 0$, $x = 2$, and $x = -4$.

27. $x^2 + x - 6 > 0$

    Factoring, we have

    $(x - 2)(x + 3) > 0$.

    We note that both factors are positive when $x > 2$ and both factors are negative when $x < -3$. Thus, x is a solution of the quadratic inequality if $x > 2$ or $x < -3$.

33. $\dfrac{x-1}{x+1} \geq 0$

    Since $\dfrac{x-1}{x+1}$ can be positive only if the factors $x - 1$ and $x + 1$ have like signs we conclude that x is a solution of the given inequality if $x \geq 1$ or $x < -1$.

37. The expression $\dfrac{1}{\sqrt{64 - 4x^2}}$ represents a real number only when the value inside the square root is positive. Thus

    $64 - 4x^2 > 0$

    or

    $64 > 4x^2$

    $16 > x^2$

We have $x^2 < 16$ which means that $-4 < x < 4$.

43. Let x be the number of students in the original group. Then each student's expenses would be $\frac{160}{x}$. If another student joins the group, the expense per student would now be $\frac{160}{x+1}$. We are told this new value is $8 less than the previous value. Thus we must solve the equation

$$\frac{160}{x} - \frac{160}{x+1} = 8.$$

Multiplying both sides of the equation by the common denominator $x(x + 1)$, we have

$160(x + 1) - 160x = 8(x)(x + 1)$
$160x + 160 - 160x = 8x^2 + 8x$

or

$0 = 8x^2 + 8x - 160.$

Dividing by 8 gives

$0 = x^2 + x - 20$
$0 = (x - 4)(x + 5).$

Thus either $x - 4 = 0$ or $x + 5 = 0$. Since x cannot be a negative number, we have $x = 4$. There were 4 students in the original group.

1 Functions

Key Ideas for Review

(This list of key ideas includes necessary concepts from Section 0.3, Straight Lines and their Equations.)

* A function is a rule or formula that determines the unique value of one variable (the dependent variable) once the value of another variable (the independent variable) has been specified.

* A single equation is not the only way to define a function. Sometimes a function is defined by a table, chart, or by several equations, in a piecewise definition.

* The domain of a function consists of the set of all real numbers at which the function is defined and yields a real number.

* The graph of the function f is the graph of the equation $y = f(x)$.

* The vertical line test: if any vertical line cuts a curve at more than one point, then the curve is not the graph of any function of x.

* The slope of a nonvertical line is given by

$$m = \frac{(y_2 - y_1)}{(x_2 - x_1)}$$

  where $P_1(x_1, y_1)$ and $P_2(x_2, y_2)$ are any two distinct points on the line.

* A vertical line has no slope. A horizontal line has 0 slope.

* If the slope m is positive, then y increases as x increases (the line *rises* from left to right); if m is negative, then y decreases as x increases (the line *falls* from left to right).

* The slope-intercept form of a line is $y = mx + b$.

1: Functions   11

* If two nonvertical lines have the same slope, then they are parallel. Conversely, if two nonvertical lines are parallel, then they have the same slope.

* The point-slope form of a line that passes through the point $P(x_1, y_1)$ and has slope m is $y - y_1 = m(x - x_1)$.

* The equation of a vertical line through (a,b) is $x = a$. The equation of a horizontal line through (a,b) is $y = b$.

* The equation of a straight line can be written as $Ax + By = C$, where A and B are not both zero. Conversely, the graph of the linear equation $Ax + By = C$ (A and B not both zero) is a straight line.

* Two lines are parallel or identical or intersect at only one point.

* The graph of a second degree function $f(x) = ax^2 + bx + c$ is a parabola, opening upward if $a > 0$ and downward if $a < 0$. The vertex has x-coordinate $x = -b/2a$. The axis of symmetry is the vertical line through the vertex.

* The parabola $y = ax^2 + bx + c$ has either a highest point (if $a < 0$) or a lowest point (if $a > 0$). This is important when we are looking for the largest or smallest value of a quadratic function.

* In business and economics problems, the cost, revenue, and profit functions occur frequently; moreover, $P(x) = R(x) - C(x)$. The "break-even" point occurs where $R(x) = C(x)$.

* The point of intersection of the graphs of two equations can be found algebraically by solving the two equations simultaneously.

* $(f \circ g)(x) = f(g(x))$.

Exercise Set 1.1 (Page 39)

3. Given the function $F(x) = \dfrac{x^2 + 1}{3x - 1}$

   (a) To find F(1) we replace x by 1. Thus

   $$F(1) = \frac{(1)^2 + 1}{3(1) - 1} = \frac{2}{2} = 1$$

   (b) To find F(-2) we replace x by -2. Thus

$$F(-2) = \frac{(-2)^2 + 1}{3(-2) - 1} = \frac{5}{-7} = \frac{-5}{7}$$

(c) To find F(4), we replace x by 4. Thus

$$F(4) = \frac{(4)^2 + 1}{3(4) - 1} = \frac{17}{11}$$

(d) To find F(0), we replace x by 0. Thus

$$F(0) = \frac{(0)^2 + 1}{3(0) - 1} = \frac{1}{-1} = -1$$

(e) To find F(a), we replace x by a. Thus

$$F(a) = \frac{(a)^2 + 1}{3(a) - 1} = \frac{a^2 + 1}{3a - 1}$$

(f) To find F(a - 2), we replace x by a - 2. Thus

$$F(a - 2)) = \frac{(a - 2)^2 + 1}{3(a - 2) - 1} = \frac{a^2 - 4a + 4 + 1}{3a - 6 - 1} = \frac{a^2 - 4a + 5}{3a - 7}$$

(g) To find F(-x), we replace x by -x Thus

$$F(-x) = \frac{(-x)^2 + 1}{3(-x) - 1} = \frac{x^2 + 1}{-3x - 1}$$

(h) To find $F(x^2)$, we replace x by x^2 Thus

$$F(x^2) = \frac{(x^2)^2 + 1}{3(x^2) - 1} = \frac{x^4 + 1}{3x^2 - 1}$$

(i)
$$\frac{1}{F(x)} = \frac{1}{\frac{x^2 + 1}{3x - 1}} = \frac{3x - 1}{x^2 + 1}$$

5. Given the formula relating Fahrenheit temperature F to Celsius temperature C

$$F = \frac{9}{5}C + 32$$

(a) To write C as a function of F, we solve the formula for C.

$$F - 32 = \frac{9}{5}C$$

or

$$C = \frac{5}{9}(F - 32)$$

(b) To find the Celsius equivalents of Fahrenheit temperatures, we substitute in the formula of part (a)

(i) If F = 4°, then $C = \frac{5}{9}(4 - 32) = \frac{5}{9}(-28) = -15.6°$

(ii) If F = 0°, then $C = \frac{5}{9}(0 - 32) = -17.8°$

(iii) If F = -10°, then $C = \frac{5}{9}(-10 - 32) = \frac{5}{9}(-42) = -23.3°$

(iv) If F = 32°, then $C = \frac{5}{9}(32 - 32) = 0°$

(v) If F = 98.6°, then $C = \frac{5}{9}(98.6 - 32) = \frac{5}{9}(66.6) = 37°$

(vi) If F = 212°, then $C = \frac{5}{9}(212 - 32) = \frac{5}{9}(180) = 100°$

9. Since a negative number has no real square root, we must have x - 1 ≥ 0 for x to be in the domain of $h(x) = \sqrt{x - 1}$. Thus, the domain of h consists of all real numbers which are greater than or equal to 1.

11. We have f(x) = (x - 2)/(x + 1). Since division by zero is undefined, f(-1) is undefined. Thus, the domain of f consists of all real numbers except -1.

19. The graph of $f(x) = 2x^2 + 3$. is the graph of $y = 2x^2 + 3$. We choose values of x arbitrarily and calculate the corresponding values of y. The results are

14 Study Guide

$y = 2x^2 + 3$

x	y
0	3
±1	5
±2	11
±3	21

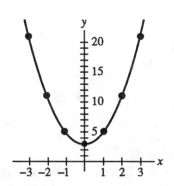

These points are plotted to obtain the graph.

23. The graph of $f(x) = \frac{1}{x}$ is the graph of $y = \frac{1}{x}$. We choose values of x arbitrarily and calculate the corresponding values for y. We note that x=0 is not in the domain of this function, so we must exclude this from our table.

$y = \frac{1}{x}$

x	y
-2	-1/2
-1	-1
-1/2	-2
1/2	2
1	1
2	1/2

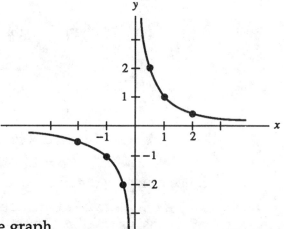

These points are plotted to obtain the graph.

Exercise Set 1.2 (Page 50)

7. Given $g(x) = 2x^2 - 8x + 9$. The axis of symmetry of the parabola is

$$x = \frac{-b}{2a} = \frac{-(-8)}{2(2)} = \frac{8}{4} = 2$$

Thus x = 2 is the x-coordinate of the vertex. The y-coordinate is

$$y = g(2) = 2(2)^2 - 8(2) + 9 = 8 - 16 + 9 = 1$$

The vertex is (2,1).

We make a table of values.

x	0	1	2	3	4
y = g(x)	9	3	1	3	9

The graph is shown below.

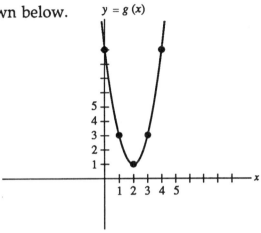

11. To find how many x-intercepts the associated graph of $f(x) = x^2 + x + 1$ has, we find the discriminant.

    Since $D = b^2 - 4ac = (1)^2 - 4(1)(1) = -3$, the discriminant is negative and the graph has no x-intercepts.

13. To find how many x-intercepts the associated graph of $f(x) = 2x^2 + 3x - 3$ has, we find the discriminant.

    Since $D = b^2 - 4ac = (3)^2 - 4(2)(-3) = 33$, the discriminant is positive and the graph has two x-intercepts.

19. (a) 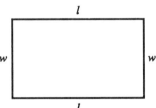 The fencing around the rectangular garden is equal to 100 feet. Thus

    $w + l + w + l = 2w + 2l = 100$
    Dividing by 2 gives $w + l = 50$ or $w = 50 - l$

    The area is given by $A = lw = l(50 - l) = 50l - l^2$

    (b) To find the length and width that will make the area as large as possible, we consider the function

    $A(l) = 50l - l^2$

Since a = -1< 0, the graph of this function is a parabola that opens downward. The maximum is attained at the vertex.

$$l = \frac{-b}{2a} = \frac{-50}{2(-1)} = 25$$

Since w = 50 - l
w = 50 - 25 = 25

and the garden should be a square to have maximum area.

23. (a) Since the bakery will sell 600 - 5x pastries daily at x cents each, the daily revenue will be

$$R(x) = x(600 - 5x)$$

Since the store will need 600 - 5x pastries daily at 50 cents each, the daily cost will be

$$C(x) = 50(600 - 5x).$$

The profit function is the revenue function minus the cost function, or

$$\begin{aligned}P(x) &= R(x) - C(x) \\ &= x(600 - 5x) - 50(600 - 5x) \\ &= (x - 50)(600 - 5x) \\ P(x) &= -5x^2 + 850x - 30{,}000\end{aligned}$$

(b) When x = 70, the profit is

$$P(70) = -5(70)^2 + 850(70) - 30{,}000 = 5{,}000 \text{ cents or } \$50.00$$

(c) Since a = -5 < 0, the graph of the profit function is a parabola that opens downward. The maximum is at the vertex.

$$x = \frac{-b}{2a} = \frac{-850}{2(-5)} = 85$$

The most profitable selling price is 85 cents.

1: Functions 17

Exercise Set 1.3 (Page 60)

3. To find the points of intersection of the graphs of $F(x) = x^2 + 3$ and $g(x) = 4x$, we equate the values of the functions by setting

    $$x^2 + 3 = 4x$$

    or

    $$x^2 - 4x + 3 = 0$$

    Factoring gives

    $$(x - 1)(x - 3) = 0$$
    $$x - 1 = 0 \quad x - 3 = 0$$
    $$x = 1 \quad x = 3$$

    Substituting these values into $F(x)$ [or $g(x)$], we have

    $$y = F(1) = 1 + 3 = 4$$
    $$y = F(3) = (3)^2 + 3 = 12$$

    Thus the points of intersection are (1,4) and (3,12).

13. To find the points of intersection of the graphs of $f(x) = 2/x$ and $g(x) = x - 1$, we equate the values of the functions by setting

    $$\frac{2}{x} = x - 1$$

    Multiplying both sides by x gives

    $$2 = x^2 - x$$

    or

    $$0 = x^2 - x - 2$$

    Factoring gives

    $$0 = (x + 1)(x - 2)$$
    $$x + 1 = 0 \quad x - 2 = 0$$
    $$x = -1 \quad x = 2$$

    Substituting these values into $f(x)$ [or $g(x)$] we have

    $$y = f(-1) = \frac{2}{-1} = -2$$

$$y = f(2) = \frac{2}{2} = 1$$

Thus the points of intersection are (-1,-2) and (2,1).

17. To find the points of intersection of the curves $x^2 + y^2 = 10$ and $9x^2 + y^2 = 18$, we shall solve the equations simultaneously.

$$x^2 + y^2 = 10$$
$$9x^2 + y^2 = 18$$

Subtracting the bottom equation from the top gives

$$\begin{array}{l} x^2 + y^2 = 10 \\ -9x^2 - y^2 = -18 \\ \hline -8x^2 = -8 \\ x^2 = 1 \\ x = \pm 1 \end{array}$$

Substituting these values into either equation gives

$$1 + y^2 = 10$$
$$y^2 = 9$$
$$y = \pm 3$$

There are four points of intersection. They are (1,3) (1,-3) (-1,3) (-1,-3).

19. To prove algebraically that the graphs of $f(x) = x^2 - x$ and $g(x) = x - 2$ do not intersect, we equate the values of functions by setting

$$x^2 - x = x - 2$$

or

$$x^2 - 2x + 2 = 0$$

The discriminant of this equation is $D = b^2 - 4ac = (-2)^2 - 4(1)(2) = 4 - 8 = -4$. Since D is negative, the equation has no real solution and there are no points of intersection.

29. The supply function is $p = S(x) = x^2 + 2x$ and the demand function is $p = D(x) = 24 - x^2$ where x is in millions of units and p is in cents per unit. To find the equilibrium price, we equate the supply and demand functions. Hence

1: Functions   19

$$S(x) = D(x)$$
$$x^2 + 2x = 24 - x^2$$

or

$$2x^2 + 2x - 24 = 0$$

Dividing the equation by 2 gives

$$x^2 + x - 12 = 0$$
$$(x - 3)(x + 4) = 0$$
$$x - 3 = 0 \quad x + 4 = 0$$
$$x = 3 \quad x = -4 \text{ (reject since } x \geq 0)$$

Thus $x = 3$ million is the number of items supplied at the equilibrium price of $p = S(3) = (3)^2 + 2(3) = 9 + 6 = 15$ cents.

Exercise Set 1.4 (Page 69)

1. The graph of $f(x) = |x| + 1$ is the graph of $y = |x| + 1$, which may also be written as

$$y = \begin{cases} x + 1 & \text{if } x \geq 0 \\ -x + 1 & \text{if } x \leq 0 \end{cases}$$

We sketch the graph in two stages. If $x \geq 0$, we have

x	0	1	2	3
f(x) = x+1	1	2	3	4

which is a linear graph in the first quadrant. If $x < 0$, we have

x	-1	-2	-3
f(x) = -x+1	2	3	4

which is a linear graph in the second quadrant. We sketch the two stages together as

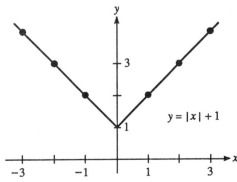

13. Given $f(x) = \dfrac{3}{x-1}$ and $g(x) = x^2 + 2$:

    (a) To find $(f \circ g)(2) = f(g(2))$, we first find $g(2)$.

    $g(2) = (2)^2 + 2 = 6$

    Thus $f(g(2)) = f(6) = \dfrac{3}{6-1} = \dfrac{3}{5}$

    (b) To find $(g \circ f)(2) = g(f(2))$, we first find $f(2)$.

    $f(2) = \dfrac{3}{2-1} = \dfrac{3}{1} = 3$

    Thus $g(f(2)) = g(3) = (3)^2 + 2 = 11$

    (c) To find $(f \circ g)(x) = f(g(x))$, we substitute $g(x)$ for every occurrence of x in the rule for f. Thus

    $f(g(x)) = \dfrac{3}{(x^2+2)-1} = \dfrac{3}{x^2+1}$

    (d) To find $(g \circ f)(x) = g(f(x))$, we substitute $f(x)$ for every occurrence of x in the rule for g. Thus

    $g(f(x)) = \left(\dfrac{3}{x-1}\right)^2 + 2 = \dfrac{9}{(x-1)^2} + 2 = \dfrac{9 + 2(x-1)^2}{x^2 - 2x + 1} = \dfrac{2x^2 - 4x + 11}{x^2 - 2x + 1}$

    (e) To find $(f \circ f)(1) = f(f(1))$, we first find $f(1)$.
    Since $x = 1$ is not in the domain of f, $f(1)$ is undefined. Thus $(f \circ f)(1)$ is undefined.

    (f) To find $(g \circ g)(2) = g(g(2))$, we first find $g(2)$.

    $g(2) = (2)^2 + 2 = 6$

    Thus $g(g(2)) = g(6) = (6)^2 + 2 = 38$.

1: Functions 21

15. given $h(x) = (5x - 3)^8$, we want to write this function as a composite of the two simpler functions. We see that $h(x) = u^8$ where $u = 5x - 3$. Thus we let $f(x) = x^8$ and $g(x) = 5x-3$. Then

$$h(x) = f(5x - 3) = f(g(x)) = (f \circ g)(x)$$

19. Given $h(x) = \left(\dfrac{3x - 5}{x + 4}\right)^{1/3}$, we want to write this function as a composite of two simpler functions. We see that $h(x) = u^{1/3}$ where $u = \dfrac{3x - 5}{x + 4}$. Thus we let $f(x) = x^{1/3}$ and $g(x) = \dfrac{3x - 5}{x + 4}$. Then $h(x) = f\left(\dfrac{3x-5}{x+4}\right) = f(g(x)) = (f \circ g)(x)$.

25. We consider three cases. Let x be the total number of people on the tour. If $x \leq 100$, then the revenue $R(x) = 300x + 300,000$. If $x \geq 150$, then the revenue $R(x) = 200x + 300,000$. If $100 < x < 150$, then the fare is decreased by $2 for each person in excess of 100 people. Thus the fare would be $300 - 2(x - 100)$ or $500 - 2x$. The revenue $R(x)$ would be $(500 - 2x)x + 300,000$. We can write the revenue as a piecewise function as follows:

$$R(x) = \begin{cases} 300x + 300,000 & \text{if } x \leq 100 \\ -2x^2 + 500x + 300,000 & \text{if } 100 < x < 150 \\ 200x + 300,000 & \text{if } x \geq 150 \end{cases}$$

Review Exercises (Page 71)

3. Since a negative number has no real square root, we must have $2x - 1 \geq 0$ for x to lie in the domain of $f(x) = \sqrt{2x - 1}$. The inequality $2x - 1 \geq 0$ is equivalent to $2x \geq 1$ and in turn $x \geq 1/2$. Thus, the domain of f consists of all real numbers which are greater than or equal to 1/2.

7. To see if the lines $x + \dfrac{1}{2}y - 4 = 0$ and $y = 8 - 2x$ are parallel, we must find their slopes.

We put $x + \dfrac{1}{2}y - 4 = 0$ into slope-intercept form.
$2x + y - 8 = 0$
$y = -2x + 8$

This gives the same equation as the first line. Hence the lines are identical and not parallel.

11. The line that goes through the points (3,2) and (-2,-3) will have a slope of

$$m = \frac{(y_2 - y_1)}{(x_2 - x_1)} = \frac{-3 - 2}{-2 - 3} = \frac{-5}{-5} = 1$$

Point-slope form is $y - y_1 = m(x - x_1)$. Using $m = 1$ and the point (3,2), we have $y - 2 = 1(x - 3)$.

13. The points (4,-2) and (-2,3) lie on the line. Let $P_1(x_1, y_1) = (4, -2)$ and $P_2(x_2, y_2) = (-2, 3)$. Using the definition of slope we have

$$m = \frac{(y_2 - y_1)}{(x_2 - x_1)} = \frac{3 - (-2)}{-2 - 4} = \frac{5}{-6} = \frac{-5}{6}$$

The point-slope equation with $m = -5/6$ and $P_1(4,-2)$ now yields

$$y - (-2) = \frac{-5}{6}(x - 4)$$

or

$$y + 2 = \frac{-5}{6}x + \frac{20}{6}$$

$$y + 2 = \frac{-5}{6}x + \frac{10}{3}$$

$$y = \frac{-5}{6}x + \frac{4}{3}.$$

15. (a) $3x^2 + y = 4$ is not a linear equation since it contains the term $3x^2$.

    (b) $2(x + 1) + \frac{1}{2}(y - 4) = 5$ is a linear equation since we could simplify this and write

$$2x + 2 + \frac{1}{2}y - 2 - 5 = 0$$

or

$$2x + \frac{1}{2}y - 5 = 0$$

which is in the form $Ax + By + C = 0$.

19. 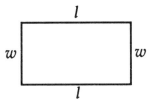 Since the wire is 20 inches, the perimeter of the rectangle is 20.

Thus $2l + 2w = 20$ or $l + w = 10$. The length, $l = 10 - w$.

The area of the rectangle is:

$$A = lw$$

or

$$A = (10 - w)w$$

The area can be written:

$$A = 10w - w^2.$$

25. The cost of $50 - 2s$ sweatshirts at $9 each is $C(s) = (50 - 2s)9 = 450 - 18s$.
The revenue obtained by selling $50 - 2s$ sweatshirts at $s each is
$R(s) = (50 - 2s)s = 50s - 2s^2$

(a) The weekly profit is $P(s) = R(s) - C(s)$
$$= 50s - 2s^2 - (450 - 18s)$$
$$P(s) = 68s - 2s^2 - 450.$$

(b) When $s = 10$, the weekly profit is

$$P(10) = 68(10) - 2(10)^2 - 450 = \$30.$$

When $s = 13$, the weekly profit is

$$P(13) = 68(13) - 2(13)^2 - 450 = \$96.$$

When $s = 20$, the weekly profit is

$$P(20) = 68(20) - 2(20)^2 - 450 = \$110.$$

(c) Since the graph of the profit function is a parabola that opens downward, the maximum profit occurs at the s-value of the vertex. Thus

$$s = \frac{-b}{2a} = \frac{-68}{2(-2)} = \frac{-68}{-4} = 17.$$

The most profitable selling price is 17 dollars.

Chapter Test (Page 73)

1. Given the function $f(x) = \frac{x-5}{3x+1}$

   To find f(0), we replace x by 0. Thus

   $$f(0) = \frac{0-5}{3(0)+1} = -5$$

   To find f(3), we replace x by 3. Thus

   $$f(3) = \frac{3-5}{3(3)+1} = \frac{-2}{10} = \frac{-1}{5}$$

   To find f(2x), we replace x by 2x. Thus

   $$f(2x) = \frac{2x-5}{3(2x)+1} = \frac{2x-5}{6x+1}$$

3. (a) Let x be the number of gallons manufactured and sold per week. Then the cost function is C(x) = 2.30x + 160 where 160 is the fixed cost and 2.30 is the variable cost. The revenue is R(x) = 3.50x. The profit P(x) is:

   $$P(x) = R(x) - C(x)$$
   $$= 3.50x - (2.30x + 160)$$
   $$P(x) = 1.20x - 160.$$

   (b) C(400) = 2.30(400) + 160 = $1080.
   R(400) = 3.50(400) = $1400.
   P(400) = 1400 - 1080 = $320.

5. The discriminant $D = b^2 - 4ac$ of the parabola $y = 3x^2 - 4x + 2$ is

$$D = (-4)^2 - 4(3)(2) = 16 - 24 = -8.$$

Since the discriminant is negative, the parabola will have no x-intercepts.

7. To find the point(s) of intersection of the graphs of $f(x) = x^2 - x$ and $g(x) = x + 3$, we equate the values of the functions by setting

$$x^2 - x = x + 3$$

or

$$x^2 - 2x - 3 = 0.$$

Factoring gives

$$(x + 1)(x - 3) = 0$$
$$x + 1 = 0 \quad x - 3 = 0$$
$$x = -1 \quad x = 3$$

Substituting these values into $f(x)$ (or $g(x)$), we have

$$y = f(-1) = 1 - (-1) = 2$$
$$y = f(3) = (3)^2 - (3) = 6$$

Thus the points of intersection are $(-1, 2)$ and $(3, 6)$.

9. The graph of

$$y = g(x) = \begin{cases} 2 + x & \text{if } x < 0 \\ 2 & \text{if } x \geq 0 \end{cases}$$

is obtained in two stages.

If $x < 0$, we have

x	-3	-2	-1
$y = 2 + x$	-1	0	1

and if $x \geq 0$, we have

x	0	1	2
$y = 2$	2	2	2

Plotting these points gives the following graph.

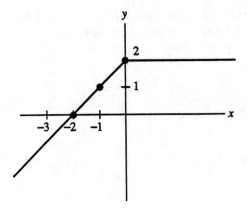

2 Limits, Continuity and Rates of Change

Key Ideas for Review

* $\lim_{x \to a} f(x) = L$ means that the values of $f(x)$ get close to the unique real number L as x gets close to (but remains different from) a.

* If the values of $f(x)$ do not approach a unique real number as x approaches the value a, we say that $\lim_{x \to a} f(x)$ does not exist.

* The limit from the left, $\lim_{x \to a^-} f(x)$, or the limit from the right, $\lim_{x \to a^+} f(x)$ may exist even when $\lim_{x \to a} f(x)$ does not exist.

* $\lim_{x \to a} f(x)$ exists and equals L only when both one-sided limits exist and equal L.

* Algebraic properties of limits.

    1. $\lim_{x \to a} k = k$.

    2. $\lim_{x \to a} x = a$.

    3a. $\lim_{x \to a} [f(x) + g(x)] = \lim_{x \to a} f(x) + \lim_{x \to a} g(x)$.

    3b. $\lim_{x \to a} [f(x) - g(x)] = \lim_{x \to a} f(x) - \lim_{x \to a} g(x)$.

4. $\lim_{x \to a} [f(x) \cdot g(x)] = \lim_{x \to a} f(x) \cdot \lim_{x \to a} g(x).$

5. $\lim_{x \to a} kf(x) = k \lim_{x \to a} f(x).$

6. $\lim_{x \to a} f(x)/g(x) = \lim_{x \to a} f(x) / \lim_{x \to a} g(x)$, if $\lim_{x \to a} g(x) \neq 0.$

7. $\lim_{x \to a} [f(x)]^{1/n} = [\lim_{x \to a} f(x)]^{1/n}$
   (where we do not take an even root of a negative number.)

* Properties 1-7 are also valid for one-sided limits.

* $+\infty$ and $-\infty$ are not numbers.

* If f(x) grows larger and larger positively (or negatively) without bound as $x \to a$, then we write $\lim_{x \to a} f(x) = +\infty$ (or $\lim_{x \to a} f(x) = -\infty$).

* Suppose that $\lim_{x \to a} g(x) = 0$ and $\lim_{x \to a} f(x) = L \neq 0.$

  1. If $f(x)/g(x) > 0$ as $x \to a$, then $\lim_{x \to a} f(x)/g(x) = +\infty.$

  2. If $f(x)/g(x) < 0$ as $x \to a$, then $\lim_{x \to a} f(x)/g(x) = -\infty.$

* A line $x = a$ is a vertical asymptote of a curve $y = f(x)$ if either $\lim_{x \to a^+} f(x) = \pm \infty$ or $\lim_{x \to a^-} f(x) = \pm \infty$ or both.

* A rational function $p(x)/q(x)$ has vertical line $x = a$ as an asymptote if $q(a) = 0$ but $p(a) \neq 0.$

* $\lim_{x \to +\infty} f(x) = L$ if the values of f(x) get closer and closer to the unique real number L as x gets larger and larger positively without bound.

* $\lim\limits_{x \to -\infty} f(x) = L$ if the values of f(x) get close to the unique real number L as x gets larger and larger negatively without bound.

* $\lim\limits_{x \to +\infty} k/x^n = 0$ and $\lim\limits_{x \to -\infty} k/x^n = 0$ for any constant k and positive integer n.

* A line y = b is a horizontal asymptote of a curve y = f(x) if $\lim\limits_{x \to +\infty} f(x) = b$, or $\lim\limits_{x \to -\infty} f(x) = b$ or both.

* f is continuous at x = a if and only if:

    1. f(a) is defined.

    2. $\lim\limits_{x \to a} f(x)$ exists.

    3. $\lim\limits_{x \to a} f(x) = f(a)$.

* If f(x) is continuous at every number in an interval I, then we say that f is continuous on the interval I. If f is continuous on $(-\infty, +\infty)$ we say that f is continuous everywhere. If f is not continuous at x = a we say that f is discontinuous there.

* Every polynomial is a continuous function everywhere.

* Every rational function is continuous everywhere except where the denominator has zero value.

* Average velocity = change in distance/change in time

* If y = f(x) then the average rate of change of y with respect to x between x_1 and x_2 is

$$\frac{y_2 - y_1}{x_2 - x_1} = \frac{f(x_2) - f(x_1)}{x_2 - x_1} = \frac{\text{change in y}}{\text{change in x}}$$

* Velocity is a rate of change.

* If y = f(x), then the instantaneous rate of change of y with respect to x at the value $x = x_0$ is:

$$\lim_{h \to 0} \frac{f(x_0 + h) - f(x_0)}{h}$$

* The instantaneous rate of change of f with respect to x at x_0 is the slope of the tangent line to the graph of f at this point $(x_0, f(x_0))$. The slope of this tangent line is called the slope of the curve $y = f(x)$ at x_0.

Exercise Set 2.1, (Page 88)

5. (a) Given

$$f(x) = \begin{cases} x & \text{if } x \leq 0 \\ x^2 & \text{if } x > 0 \end{cases}$$

We complete the tables.

x	1	.5	.1	.01	.001
$f(x) = x^2$	1	.25	.01	.0001	.00001

x	-1	-.5	-.1	-.01	-.001
$f(x) = x$	-1	-.5	-.1	-.01	-.001

All values of x are positive in the table on top. Thus $f(x) = x^2$.
All values of x are negative in the bottom table. Thus $f(x) = x$.

To find $\lim_{x \to 0^-} f(x)$, we use the bottom table. $\lim_{x \to 0^-} f(x) = 0$.

To find $\lim_{x \to 0^+} f(x)$, we use the top table. $\lim_{x \to 0^+} f(x) = 0$.

Since $\lim_{x \to 0^-} f(x) = \lim_{x \to 0^+} f(x) = 0$, we have $\lim_{x \to 0} f(x) = 0$.

(b)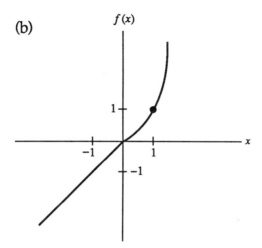

9. Since f(x) = 3x + 4 is a polynomial, we have
   $\lim_{x \to -2} (3x + 4) = 3(-2) + 4 = -2$

11. Since p(x) = $3x^2$ + 2x -5 is a polynomial, we have
    $\lim_{x \to 2} (3x^2 + 2x - 5) = 3(2)^2 + 2(2) - 5 = 11$.

13. Since $\lim_{x \to 4} (x^2 + 9) \neq 0$, we can use the quotient property of limits.

    Thus

    $$\lim_{x \to 4} \frac{2x+5}{x^2+9} = \lim_{x \to 4} 2x+5 / \lim_{x \to 4} x^2 + 9 = \frac{2(4)+5}{(4)^2+9} = \frac{13}{25}$$

17. Since $\lim_{x \to 0} x = 0$ in the denominator, we cannot use the limit of a quotient property (Property 6). But, the limit of the numerator is also zero. Thus, we factor the numerator, obtaining

    $$\lim_{x \to 0} \frac{x^2 - 2x}{x} = \lim_{x \to 0} \frac{x(x-2)}{x} = \lim_{x \to 0} (x-2) = -2$$

19. Given $\lim_{x \to 4} \frac{x+1}{x^2 - 3x - 4}$

We examine the limits of both the numerator and the denominator. Since $\lim_{x \to 4} x^2 - 3x - 4 = 0$ and $\lim_{x \to 4} x + 1 \neq 0$, the given limit does not exist.

27. Given $\lim_{x \to 0} \sqrt[4]{x}$

    From property 7, this limit does not exist, since $n = 4$ is even and $a = 0$.

35. Given $\lim_{h \to 0} \dfrac{(x+h)^2 - x^2}{h}$

    $$\lim_{h \to 0} \dfrac{(x+h)^2 - x^2}{h} = \lim_{h \to 0} \dfrac{(x^2 + 2xh + h^2 - x^2)}{h} = \lim_{h \to 0} \dfrac{2xh + h^2}{h}$$

    $$= \lim_{h \to 0} (2x + h) = 2x$$

    Since $2x$ does not involve h, it acts as a constant with respect to h.

37. (a) If x is to the left of 2, $f(x) = 2x + 1$. Thus, as x approaches 2 from the left, the corresponding $f(x)$ values approach $2(2) + 1 = 5$. If x is to the right of 2, $f(x) = -5x/2 + 10$. Thus, as x approaches 2 from the right, the corresponding $f(x)$ values approach $-5(2)/2 + 10 = 5$. Since as x approaches 2 the values of $f(x)$ get closer and closer to the number 5, we write $\lim_{x \to 2} f(x) = 5$.

    (b)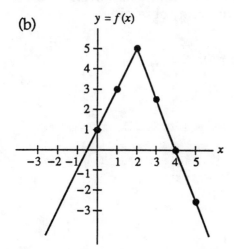

x	f(x) = 2x+1	x	f(x) = -5x/2 + 10
0	1	3	2.5
1	3	4	0
2	5	5	-2.5

39. Given $f(x) = \dfrac{x^2 - 25}{x + 5}$

$$\lim_{x \to -5^+} \frac{x^2 - 25}{x + 5} = \lim_{x \to -5^+} \frac{(x+5)(x-5)}{x+5} = -10$$

$$\lim_{x \to -5^-} \frac{x^2 - 25}{x + 5} = \lim_{x \to -5^-} \frac{(x+5)(x-5)}{x+5} = -10$$

Thus since $\lim_{x \to -5^+} f(x) = \lim_{x \to -5^-} f(x) = -10$, we have $\lim_{x \to -5} f(x) = -10$.

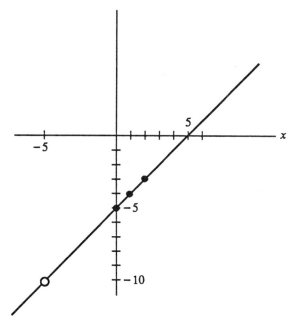

45. As x approaches 3 from the right, the values of f(x) approach 6. As x approaches 3 from the left, the values of f(x) approach 3. Hence, the values of f(x) do not approach a single finite number as x approaches 3, so that $\lim_{x \to 3} f(x)$ does not exist.

34 Study Guide

Exercise Set 2.2, (Page 101)

5. $\lim\limits_{x \to 2^+} \dfrac{x}{x-2} = +\infty$ because

   $\lim\limits_{x \to 2^+} (x-2) = 0$, $\lim\limits_{x \to 2^+} x = 2 \neq 0$, and $\dfrac{x}{x-2} = \dfrac{+}{+} > 0$ as $x \to 2^+$ ($x > 2$)

7. $\lim\limits_{x \to 3^-} \dfrac{1}{x^2-9} = -\infty$ because

   $\lim\limits_{x \to 3^-} (x^2 - 9) = 0$, $\lim\limits_{x \to 3^-} 1 = 1 \neq 0$ and $\dfrac{1}{x^2-9} = \dfrac{+}{-} < 0$ as $x \to 3^-$ ($x < 3$)

13. $\lim\limits_{x \to 2} \dfrac{x-2}{x^2-x-2}$

    Since $\lim\limits_{x \to 2} (x-2) = 0$ and $\lim\limits_{x \to 2} (x^2 - x - 2) = 0$, we factor the denominator

    Thus

    $$\lim\limits_{x \to 2} \dfrac{x-2}{x^2-x-2} = \lim\limits_{x \to 2} \dfrac{(x-2)}{(x+1)(x-2)} = \lim\limits_{x \to 2} \dfrac{1}{x+1} = \dfrac{1}{3}$$

19. $\lim\limits_{x \to +\infty} \dfrac{1}{x+1}$

    As x gets larger and larger positively without bound, the denominator gets larger and larger positively without bound. Hence $\lim\limits_{x \to +\infty} \dfrac{1}{x+1} = 0$.

23. $\lim\limits_{x \to +\infty} (3 + 4x)$

As x gets larger and larger positively without bound, so does 3 + 4x. Hence $\lim\limits_{x \to +\infty} (3 + 4x) = +\infty$.

25. $\lim\limits_{x \to +\infty} \dfrac{5x}{x+3}$

In this case, the limits of the numerator and denominator do not exist. If we divide numerator and denominator by the highest power of x in the denominator, we have

$$\lim_{x \to +\infty} \frac{5x}{x+3} = \lim_{x \to +\infty} \frac{\frac{5x}{x}}{\frac{x}{x}+\frac{3}{x}} = \lim_{x \to +\infty} \frac{5}{1+\frac{3}{x}} = 5$$

since $\lim\limits_{x} \dfrac{3}{x} = 0$ from Property 9.

29. $\lim\limits_{x \to -\infty} \dfrac{x^3}{x^2+7}$

In this case, the limits of the numerator and the denominator do not exist. We write $\dfrac{x^3}{x^2+7}$ as the product $x \left(\dfrac{x^2}{x^2+7} \right)$. Thus

$$\lim_{x \to -\infty} \frac{x^3}{x^2+7} = [\lim_{x \to -\infty} x\,][\lim_{x \to -\infty} \frac{x^2}{x^2+7}] = [\lim_{x \to -\infty} x][\lim_{x \to -\infty} \frac{\frac{x^2}{x^2}}{\frac{x^2}{x^2}+\frac{7}{x^2}}] =$$

$$[\lim_{x \to -\infty} x\,][\lim_{x \to -\infty} \frac{1}{1+\frac{7}{x^2}}\,][\lim_{x \to -\infty} x][1] = -\infty$$

31. In this case, the limits of the numerator and denominator do not exist. If we divide the numerator and denominator by x^2 (the highest power of x in the denominator), we obtain

$$\lim_{x \to +\infty} \frac{2x^2 + 5x - 3}{5x^2 - 3x + 1} = \lim_{x \to +\infty} \frac{\frac{2x^2}{x^2} + \frac{5x}{x^2} - \frac{3}{x^2}}{\frac{5x^2}{x^2} - \frac{3x}{x^2} + \frac{1}{x^2}} = \lim_{x \to +\infty} \frac{2 + \frac{5}{x} - \frac{3}{x^2}}{5 - \frac{3}{x} + \frac{1}{x^2}} = \frac{2}{5}$$

33. In this case, the limits of the numerator and denominator do not exist. If we divide the numerator and denominator by x^2 (the highest power of x in the denominator), we obtain

$$\lim_{x \to +\infty} \frac{2x}{x^2 - 3x} = \lim_{x \to +\infty} \frac{\frac{2x}{x^2}}{\frac{x^2}{x^2} - \frac{3x}{x^2}} = \lim_{x \to +\infty} \frac{\frac{2}{x}}{1 - \frac{3}{x}} = \frac{0}{1} = 0$$

35. As x gets larger and larger without bound, so does $3x^2 + 2$, so $\lim_{x \to -\infty} (3x^2 + 2)$ does not exist.

    We write $\lim_{x \to -\infty} (3x^2 + 2) = +\infty$

37. Given $y = \frac{x+1}{x+2}$

    The line $x = -2$ is a vertical asymptote of the curve since

    $$\lim_{x \to -2^+} \frac{x+1}{x+2} = -\infty \text{ and } \lim_{x \to -2^-} \frac{x+1}{x+2} = +\infty$$

    The line $y = 1$ is a horizontal asymptote of the curve, since

$$\lim_{x \to \pm\infty} \frac{x+1}{x+2} = \lim_{x \to \pm\infty} \frac{\frac{x}{x} + \frac{1}{x}}{\frac{x}{x} + \frac{2}{x}} = \lim_{x \to \pm\infty} \frac{1 + \frac{1}{x}}{1 + \frac{2}{x}} = \frac{1+0}{1+0} = 1$$

Exercise Set 2.3, (Page 108)

3. (a) Note that f(6) is not defined and thus f(x) is discontinuous at x = 6. When x ≠ 6, f(x) = (x-6)(x+6)/(x-6) = x + 6 so that we obtain the graph

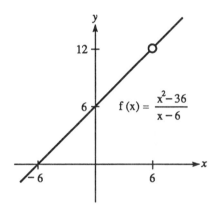

Note also that the graph as a hole at x = 6 and $\lim_{x \to 6} f(x) = 12$.

(b) When x = 6, f(6) is defined, in fact, f(6) = 12. Also,

$$\lim_{x \to 6} f(x) = \lim_{x \to 6} \frac{(x-6)(x+6)}{x-6} = \lim_{x \to 6} (x+6) = 12 = f(6)$$

We conclude that f(x) is continuous at x = 6. Note that for all x, the rule for the function f can be simplified to f(x) = x + 6. Thus the graph of f is the graph of the line y = x + 6.

(c) The rule for f can be simplified to

$$f(x) = \begin{cases} x + 6 & \text{if } x \neq 6 \\ 8 & \text{if } x = 6 \end{cases}$$

Thus, f(6) = 8 and $\lim_{x \to 6} f(x) = \lim_{x \to 6} (x + 6) = 12 \neq f(6)$ so that f is discontinuous at x = 6. The graph of f is

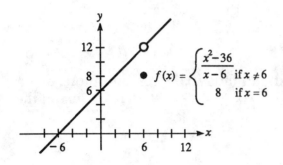

7. Since f(3) is undefined, f(x) is discontinuous at x = 3.

9. Note that f(2) = 5 and that $\lim_{x \to 2} f(x) = \lim_{x \to 2} (2x) = 2(2) = 4 \neq f(2)$.
Thus, f(x) is discontinuous at x = 2.

13. As a rational function f(x) is continuous at all values of x for which the denominator is nonzero. The denominator is zero only at x = 3. Thus, f(x) is continuous everywhere except at x = 3.

17. If a < 1, $\lim_{x \to a} f(x) = \lim_{x \to a} 3x = 3a = f(a)$, so f is continuous at values less than 1. If a > 1, $\lim_{x \to a} f(x) = \lim_{x \to a} (2x + 1) = 2a + 1 = f(a)$, so f is continuous at values greater than 1. If a = 1, f(a) = f(1) = 3(1) = 3. Also, as x approaches 1 from the right, the values of f(x) approach 2(1) + 1 = 3. As x approaches 1 from the left, the values of f(x) approach 3(1) = 3. Thus, $\lim_{x \to 1} f(x) = 3 = f(1)$ so that f is continuous at x = 1. In summary, f is continuous everywhere.

19. Since $f(x) = \sqrt{x+2}$ is the square root of a continuous function, it is continuous at values of x where we take the square root of a non-negative number (see Property 3). Thus f(x) is continuous when $x + 2 > 0$ or $x > -2$.

    f(x) is not continuous at x = -2 since $\lim_{x \to -2^-} f(x)$ does not exist.

27. Since $t^2 - 5t + 6 = (t-3)(t-2)$, the rule for M can be written as $M(t) = (t-3)(t-2)/(t-3) = t-2$ when $t \neq 3$. Thus, M is continuous for $t \neq 3$. Also $\lim_{t \to 3} M(t) = \lim_{t \to 3} (t-2) = 1$.

    If M(3) is assigned value 1, then $M(3) = \lim_{t \to 3} M(t)$ and the redefined function M is also continuous at t = 3.

Exercise Set 2.4, (Page 121)

5. The average rate of change of y = f(x) with respect to x is

    $$\frac{f(x_2) - f(x_1)}{x_2 - x_1}$$

    Given $y = f(x) = 2x^2 + 3$, $x_1 = -2$, and $x_2 = 5$. The average rate of change is

    $$\frac{f(5) - f(-2)}{5-(-2)} = \frac{2(5)^2 + 3 - [2(-2)^2 + 3]}{5-(-2)} = \frac{53-11}{7} = \frac{42}{7} = 6$$

9. (a) The average rate of change of y = f(x) with respect to x between $x_1 = 0$ and $x_2 = 3$ is $\frac{f(x_2) - f(x_1)}{x_2 - x_1}$. Here $f(x) = x^2 - 2x$, so that $f(x_2) = 3^2 - 2(3) = 3$ and $f(x_1) = 0^2 - 2(0) = 0$. Thus, the desired average rate of change is $\frac{3-0}{3-0} = 1$.

    (b) The graph of $y = x^2 - 2x$ is obtained by selecting values of x arbitrarily (but centered at the turning point x = 1 of the parabola) and calculating the corresponding y values. The results are

x	-2	-1	0	1	2	3	4
$y = x^2 - 2x$	8	3	0	-1	0	3	8

These points are plotted to obtain the graph. The secant line joining the points (0,0) and (3,3) from part (a) is also sketched

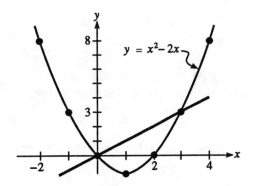

(c) Here $f(x_2) = f(1+h) = (1+h)^2 - 2(1+h)$

$$= 1 + 2h + h^2 - 2 - 2h$$

$$= h^2 - 1$$

and $f(x_1) = f(1) = (1)^2 - 2(1) = -1$. Thus, the desired average rate of change is

$$\frac{f(x_2) - f(x_1)}{x_2 - x_1} = \frac{(h^2 - 1) - (-1)}{(1+h) - 1} = \frac{h^2}{h} = h$$

(d) The instantaneous rate of change of y with respect to x at the value $x = x_0$ is given by

$$\lim_{h \to 0} \frac{f(x_0 + h) - f(x_0)}{h}$$

Since $x_0 = 1$, we have

$$\lim_{h \to 0} \frac{f(1+h) - f(1)}{h} = \lim_{h \to 0} \frac{(1+h)^2 - 2(1+h) - (-1)}{h} =$$

$$\lim_{h \to 0} \frac{1 + 2h + h^2 - 2 - 2h + 1}{h} = \lim_{h \to 0} \frac{h^2}{h} = \lim_{h \to 0} h = 0.$$

(e) The instantaneous rate of change at x_0 is the slope of the tangent line to the graph of f at the point $(x_0, f(x_0))$. Thus from (d), the slope of the tangent line to the curve $y = x^2 - 2x$ at $(1, -1)$ is equal to 0.

11. Given $y = x^2 - 2x$. The instantaneous rate of change of y with respect to x at $x = x_0$ is

$$\lim_{h \to 0} \frac{(x_0 + h)^2 - 2(x_0 + h) - (x_0^2 - 2x_0)}{h} =$$

$$\lim_{h \to 0} \frac{x_0^2 - 2x_0 h + h^2 - 2x_0 - 2h - x_0^2 + 2x_0}{h} =$$

$$\lim_{h \to 0} \frac{2x_0 h + h^2 - 2h}{h} = \lim_{h \to 0} (2x_0 + h - 2) = 2x_0 - 2$$

17. The instantaneous rate of change of f(x) at x_0 is $\lim_{h \to 0} \frac{f(x_0 + h) - f(x_0)}{h}$.

Here $f(x) = x^2$ and $x_0 = 1$ so that the desired rate of change is

$$\lim_{h \to 0} \frac{f(1 + h) - f(1)}{h} = \lim_{h \to 0} \frac{(1 + h)^2 - (1)^2}{h} =$$

$$\lim_{h \to 0} \frac{1 + 2h + h^2 - 1}{h} = \lim_{h \to 0} \frac{2h + h^2}{h} = \lim_{h \to 0} 2 + h = 2.$$

21. Given the cost function $C(x) = x^2 - 4x + 8$

(a) The average rate of change of cost with respect to the number of gallons produced is $[C(x_2) - C(x_1)]/(x_2 - x_1)$. Here $C(x) = x^2 - 4x + 8$ so that $C(x_1) = C(1/2) = (1/2)^2 - 4(1/2) + 8 = 6.25$ and $C(x_2) = C(2) = (2)^2 - 4(2) + 8 = 4$ million dollars. Thus, the desired average rate of change is $(4 - 6.25)/(2 - .5) = -1.5$ million dollars per million gallons produced, that is-- $1.50 per gallon.

(b) The average rate of change of cost when the level of production changes from $x_1 = 3$ to $x_2 = 5$ is

$$\frac{C(5) - C(3)}{5-3} = \frac{(5)^2 - 4(5) + 8 - [(3)^2 - 4(3) + 8]}{5-3}$$

$$= \frac{25 - 20 + 8 - 9 + 12 - 8}{2} = 4$$

Thus the average rate of change is $4 per gallon.

(c) The instantaneous rate of change of cost at x_0 is
$\lim\limits_{h \to 0} \frac{C(x_0 + h) - C(x_0)}{h}$. With $x_0 = 1$ and $C(x) = x^2 - 4x + 8$ we obtain

$$\lim_{h \to 0} \frac{C(1 + h) - C(1)}{h} =$$

$$\lim_{h \to 0} \frac{(1 + h)^2 - 4(1 + h) + 8 - [(1)^2 - 4(1) + 8]}{h} =$$

$$\lim_{h \to 0} \frac{[1 + 2h + h^2 - 4 - 4h + 8] - [5]}{h} = \lim_{h \to 0} \frac{h^2 - 2h}{h} = \lim_{h \to 0} h - 2 = -2$$

Thus, the cost is decreasing at $2.00 per gallon instantaneous rate when the level of production is $x_0 = 1$ million gallons.

(d) The instantaneous rate of change of cost when $x_0 = 2$ is

$$\lim_{h \to 0} \frac{C(2 + h) - C(2)}{h} =$$

$$\lim_{h \to 0} \frac{(2 + h)^2 - 4(2 + h) + 8 - [(2)^2 - 4(2) + 8]}{h} =$$

$$\lim_{h \to 0} \frac{[4 + 4h + h^2 - 8 - 4h + 8] - [4]}{h} = \lim_{h \to 0} h = 0$$

Thus, the cost doesn't change when the level of production is $x_0 = 2$ million gallons.

(e) The instantaneous rate of change of cost when $x_0 = 3$ is

$$\lim_{h \to 0} \frac{C(3+h) - C(3)}{h} =$$

$$\lim_{h \to 0} \frac{(3+h)^2 - 4(3+h) + 8 - [(3)^2 - 4(3) + 8]}{h} =$$

$$\lim_{h \to 0} \frac{9 + 6h + h^2 - 12 - 4h + 8 - 5}{h} = \lim_{h \to 0} \frac{2h + h^2}{h} =$$

$$\lim_{h \to 0} (2 + h) = 2$$

Thus, the cost is increasing at the rate of \$2 per gallon when the level of production is $x_0 = 3$ million gallons.

23. Given $d = 2t + 6t^2$. The average velocity is given by $[d(t_2) - d(t_1)]/(t_2 - t_1)$.

(a) If $t_1 = 0$ and $t_2 = 4$, then the average velocity is

$$\frac{d(4) - d(0)}{4 - 0} = \frac{2(4) + 6(4)^2 - 0}{4} = \frac{8 + 96}{4} = \frac{104}{4} = 26 \text{ miles per hour}$$

(b) If $t_1 = 2$ and $t_2 = 5$, then the average velocity is

$$\frac{d(5) - d(2)}{5 - 2} = \frac{2(5) + 6(5)^2 - [2(2) + 6(2)^2]}{5 - 2} = \frac{10 + 150 - (4 + 24)}{3} = \frac{132}{3}$$

$$= 44 \text{ miles per hour}$$

(c) Since the distance d is given by $d = f(t) = 2t + 6t^2$, the instantaneous velocity at $t_0 = 3$ is

44 Study Guide

$$\lim_{h \to 0} \frac{[2(3+h) + 6(3+h)^2] - [2(3) + 6(3)^2]}{h} =$$

$$\lim_{h \to 0} \frac{[6 + 2h + 6(9 + 6h + h^2)] - [6 + 54]}{h} =$$

$$\lim_{h \to 0} \frac{[60 + 38h + 6h^2] - 60}{h} = \lim_{h \to 0} \frac{38h + 6h^2}{h} = \lim_{h \to 0} (38 + 6h) = 38$$

Thus the instantaneous velocity at $t_0 = 3$ is 38 miles per hour.

Review Exercises, (Page 124)

5. Since the limit of both the denominator and numerator is zero, we factor. Thus

$$\lim_{x \to 8} \frac{64 - x^2}{8 - x} = \lim_{x \to 8} \frac{(8-x)(8+x)}{(8-x)} = \lim_{x \to 8} (8 + x) = 16$$

7. In this case, the limits of the numerator and the denominator do not exist. But if we divide the numerator and denominator by x (the highest power of x in the denominator), we obtain

$$\lim_{x \to +\infty} \frac{4 - 2x}{3 + 5x} = \lim_{x \to +\infty} \frac{\frac{4}{x} - \frac{2x}{x}}{\frac{3}{x} + \frac{5x}{x}} = \lim_{x \to +\infty} \frac{\frac{4}{x} - 2}{\frac{3}{x} + 5} = \frac{-2}{5}$$

13. Given $f(x) = \frac{x^2 - 3x + 10}{x + 2}, x \neq -2$

    To make f(x) continuous everywhere, we must define $f(-2) = \lim_{x \to -2} f(x)$. Since

$$\lim_{x \to -2} \frac{x^2 - 3x + 10}{x + 2} = \lim_{x \to -2} \frac{(x+2)(x-5)}{x+2} = \lim_{x \to -2} (x - 5) = -7$$

We must define $f(-2) = -7$

15. Note that $f(-2)$ is defined, in fact, $f(-2) = -1$. Also

$$\lim_{x \to -2} \frac{x^2 - 3x + 2}{x + 2} = \lim_{x \to -2} \frac{(x+1)(x+2)}{(x+2)} = \lim_{x \to -2} (x+1) = -1$$

Thus, $f(-2) = -1 = \lim_{x \to -2} f(x)$ so that f is continuous at $x = -2$. Since, for all x the rule for $f(x)$ can be simplified to $f(x) = x + 1$, the graph of f is the graph of the line $y = x + 1$.

19. $f(x) = 1/(x+1)$
$f(x)$ is not continuous at $x = -1$, since $f(-1)$ is not defined.

23. As a rational function, $f(x) = (x-1)/x$ is continuous at all values of x for which the denominator is nonzero. The denominator is zero only when $x = 0$. Hence, f is continuous everywhere except at $x = 0$.

25. If $y = f(x) = 2x^2 - 1$, the average rate of change of y with respect to x between $x_1 = -2$ and $x_2 = 3$ is

$$\frac{f(x_2) - f(x_1)}{x_2 - x_1} = \frac{f(3) - f(-2)}{3 - (-2)} =$$

$$\frac{2(3)^2 - 1 - [2(-2)^2 - 1]}{3 - (-2)} = \frac{18 - 1 - 8 + 1}{5} = \frac{10}{5} = 2$$

27. The instantaneous rate of change of $y = f(x)$ at x_0 is $\lim_{h \to 0} \frac{f(x_0 + h) - f(x_0)}{h}$. When $x_0 = 3$ and $f(x) = 2x^2 + 2$, we obtain

$$\lim_{h \to 0} \frac{f(3 + h) - f(3)}{h} = \lim_{h \to 0} \frac{[2(3+h)^2 + 2] - [2(3)^2 + 2]}{h} =$$

$$\lim_{h \to 0} \frac{f(3+h) - f(3)}{h} = \lim_{h \to 0} \frac{[2(3+h)^2 + 2] - [2(3)^2 + 2]}{h} =$$

$$\lim_{h \to 0} \frac{[2(9 + 6h + h^2) + 2] - [18 + 2]}{h} =$$

$$\lim_{h \to 0} \frac{20 + 12h + 2h^2 - 20}{h} = \lim_{h \to 0} \frac{12h + 2h^2}{h} = \lim_{h \to 0} (12 + 2h) = 12$$

31. The weight function is $W(t) = 3t^2$ for $0 \leq t \leq 39$

    (a) The average rate of change in the weight between $t_1 = 4$ and $t_2 = 10$ is

    $$\frac{W(t_2) - W(t_1)}{t_2 - t_1} = \frac{W(10) - W(4)}{10 - 4} = \frac{3(10)^2 - 3(4)^2}{10 - 4} = \frac{300 - 48}{6}$$

    $$= \frac{252}{6} = 42 \text{ grams per week.}$$

    (b) The instantaneous rate of change at $t_0 = 20$ is

    $$\lim_{h \to 0} \frac{W(t_0 + h) - W(t_0)}{h} = \lim_{h \to 0} \frac{W(20 + h) - W(20)}{h} =$$

    $$\lim_{h \to 0} \frac{3(20 + h)^2 - 3(20)^2}{h} = \lim_{h \to 0} \frac{1200 + 120h + 3h^2 - 1200}{h} =$$

    $$\lim_{h \to 0} \frac{120h + 3h^2}{h} = \lim_{h \to 0} (120 + 3h) = 120$$

    Thus, when $t_0 = 20$, the instantaneous rate of change of the weight of the fetus is 120 grams per week.

2 : Limits, Continuity... 47

Chapter Test, (Page 127)

1. (a) Since $\lim_{x \to 3} (x - 1) \neq 0$, we can use the quotient property of limits. Thus

    $$\lim_{x \to 3} (x + 2)/(x - 1) = \lim_{x \to 3} (x + 2) / \lim_{x \to 3} (x - 1) = (3 + 2)/(3 - 1) = 5/2.$$

    (b) Since $\lim_{x \to 1} (x - 1) = 0$ in the denominator, we cannot use the quotient property. But, the limit of the numerator is also zero. Thus, we factor the numerator, and we obtain

    $$\lim_{x \to 1} \frac{x^2 - 1}{x - 1} = \lim_{x \to 1} \frac{(x - 1)(x + 1)}{(x - 1)} = \lim_{x \to 1} (x + 1) = 2$$

3. (a) We divide the numerator and the denominator by the highest power of x that appears in the denominator. Thus

    $$\lim_{x \to +\infty} \frac{3x}{1 - 6x} = \lim_{x \to +\infty} \frac{\frac{3x}{x}}{\frac{1}{x} - \frac{6x}{x}} = \lim_{x \to +\infty} \frac{3}{\frac{1}{x} - 6} = \frac{3}{-6} = \frac{-1}{2}$$

    since $\lim_{x \to +\infty} \frac{1}{x} = 0$.

    (b) We divide the numerator and the denominator by x^2 since it is the highest power of x that appears in the denominator. Thus

    $$\lim_{x \to -\infty} \frac{2x}{x^2 - 3} = \lim_{x \to -\infty} \frac{\frac{2x}{x^2}}{\frac{x^2}{x^2} - \frac{3}{x^2}} = \lim_{x \to -\infty} \frac{\frac{2}{x}}{1 - \frac{3}{x^2}} = \frac{0}{1-0} = 0$$

since $\lim_{x \to -\infty} \frac{2}{x} = 0$ and $\lim_{x \to -\infty} \frac{3}{x^2} = 0$.

5. (a) The function is continuous everywhere.
   (b) The function is not continuous at x = a, since f(a) is not defined.
   (c) The function is not continuous at x = c, since $\lim_{x \to c^+} f(x) \neq \lim_{x \to c^-} f(x)$

   (d) The function is not continuous at x = a, since $\lim_{x \to a} f(x) \neq f(a)$.

7. The function $f(x) = \frac{1}{2x^2 - 7x - 4}$ is continuous at all values of x except those for which the denominator is equal to zero. Setting

    $2x^2 - 7x - 4 = 0$
    $(2x + 1)(x - 4) = 0$
    $x = -1/2 \quad x = 4$

    At x = -1/2 and at x = 4, f(x) is not defined, so it is not continuous there.

9. Given $f(x) = x^2 - 3x$

    (a) The average rate of change in y with respect to x, from $x_1 = 1$ to $x_2 = 4$ is

    $$\frac{f(x_2) - f(x_1)}{x_2 - x_1} = \frac{f(4) - f(1)}{4-1} = \frac{(4)^2 - 3(4) - [(1)^2 - 3(1)]}{4-1}$$

    $$= \frac{16 - 12 - 1 + 3}{3} = \frac{6}{3} = 2$$

    (b) We make a table. Since -b/2a = 3/2, the vertex of this parabola will be at 1.5

x	0	1	1.5	2	3	4
$f(x) = x^2 - 3x$	0	-2	-2.25	-2	0	4

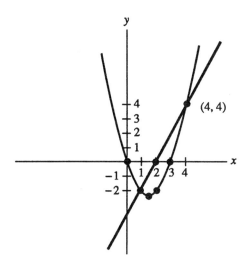

(c) The average rate of change in y with respect to x from $x_1 = 1$ to $x_2 = 1 + h$ is

$$\frac{f(x_2) - f(x_1)}{x_2 - x_1} = \frac{f(1+h) - f(1)}{(1+h) - 1} = \frac{(1+h)^2 - 3(1+h) - [(1)^2 - 3(1)]}{h}$$

$$= \frac{1 + 2h + h^2 - 3 - 3h - 1 + 3}{h} = \frac{h^2 - h}{h} = h - 1$$

(d) The instantaneous rate of change of y with respect to x at $x_0 = 1$ is

$$\lim_{h \to 0} \frac{f(1+h) - f(1)}{h} = \lim_{h \to 0} \frac{(1+h)^2 - 3(1+h) - [(1)^2 - 3(1)]}{h}$$

$$= \lim_{h \to 0} \frac{1 + 2h + h^2 - 3 - 3h - 1 + 3}{h} = \lim_{h \to 0} \frac{h^2 - h}{h} = \lim_{h \to 0} h - 1 = -1$$

3 The Derivative

Key Ideas for Review

* The derivative of the function f at x_0 is defined by

$$f'(x_0) = \lim_{h \to 0} \frac{f(x_0 + h) - f(x_0)}{h}$$

   if this limit exists.

* If $f'(x_0)$ exists, then we say that f is differentiable at x_0.

* $f'(x_0)$ is the slope of the tangent line to the curve $y = f(x)$ at the point where $x = x_0$.

* The equation of the tangent line to the graph of f at the point $(x_0, f(x_0))$ is

$$y - f(x_0) = f'(x_0)(x - x_0).$$

* If f is differentiable at x_0, then f must be continuous at x_0. However, we have seen examples of functions that are continuous at x_0 but not differentiable there.

* Alternate definition of the derivative

$$f'(x_0) = \lim_{x \to x_0} \frac{f(x) - f(x_0)}{x - x_0}.$$

* Delta notation: If $y = f(x)$, then Δx = a change in x, while

$$\Delta y = \text{the corresponding change in } y = f(x + \Delta x) - f(x).$$

   Then

$$\frac{dy}{dx} = \lim_{\Delta x \to 0} \frac{\Delta y}{\Delta x}$$

* Alternative notation for derivatives:

$$\frac{dy}{dx} = \frac{d}{dx}(y) = D_x(y) = y;$$

$$\frac{d}{dx}[f(x)] = D_x[f(x)] = f'(x).$$

* If f is a constant function, then $f'(x) = 0$.

* For any rational number r, $\frac{d}{dx}[x^r] = rx^{r-1}$.

* For any constant, k, $\frac{d}{dx}[kf(x)] = k\frac{d}{dx}[f(x)]$.

* $\frac{d}{dx}[f(x) + g(x)] = \frac{d}{dx}[f(x)] + \frac{d}{dx}[g(x)]$.

* $\frac{d}{dx}[f(x) - g(x)] = \frac{d}{dx}[f(x)] - \frac{d}{dx}[g(x)]$.

* $\frac{d}{dx}[f(x)g(x)] = f(x)\frac{d}{dx}g(x) + g(x)\frac{d}{dx}f(x)$.

* $\frac{d}{dx}[\frac{f(x)}{g(x)}] = \frac{g(x)f'(x) - f(x)g'(x)}{[g(x)]^2}$

* When x items have been made, the marginal cost C'(x) is the rate of change of the total cost per unit change in the level of production. Similar remarks are true for the marginal revenue R'(x) and marginal profit P'(x).

* The cost of making item (x + 1) is approximately C'(x). Similar interpretations of R'(x) and P'(x) are valid.

* The chain rule for composite functions: If $y = f(u)$ and $u = g(x)$ then

$$\frac{dy}{dx} = \frac{dy}{du} \cdot \frac{du}{dx}$$

* For any rational number r, $\frac{d}{dx}[g(x)]^r = r[g(x)]^{r-1}g'(x)$.

*   To solve related rates problems, first relate the variables by an equation, then differentiate both sides using the chain rule, and finally substitute in constants.

*   Given an equation that defines y implicitly as a function of x, dy/dx can be obtained from that equation by implicit differentiation.

*   Implicit differentiation is often useful in related rates problems.

Exercise Set 3.1, (Page 138)

5. Given $f(x) = 3x^2 + 2$

    We first find $f(x+h) = 3(x+h)^2 + 2 = 3x^2 + 6xh + 3h^2 + 2$

    We then find $f(x+h) - f(x) = (3x^2 + 6xh + 3h^2 + 2) - (3x^2 + 2)$
    $$= 6xh + 3h^2$$

    The difference quotient is

    $$\frac{f(x+h) - f(x)}{h} = \frac{6xh + 3h^2}{h} = 6x + 3h$$

11. Given $f(x) = x^2 - 2x + 4$

    (a) To find f'(x) by the four-step procedure we first find f(x+h) by replacing every occurrence of x in the rule for f(x) by x+h. Thus

    Step 1: $f(x+h) = (x+h)^2 - 2(x+h) + 4 = x^2 + 2xh + h^2 - 2x - 2h + 4$

    Step 2: $f(x+h) - f(x) = (x^2 + 2xh + h^2 - 2x - 2h + 4) - (x^2 - 2x + 4)$
    $$= 2xh + h^2 - 2h$$

    Step 3: $\dfrac{f(x+h) - f(x)}{h} = \dfrac{2xh + h^2 - 2h}{h} = 2x + h - 2$

    Step 4: $\lim\limits_{h \to 0} \dfrac{f(x+h) - f(x)}{h} = \lim\limits_{h \to 0} (2x + h - 2) = 2x - 2$

    Thus $f'(x) = 2x - 2$

    (b) We substitute x = 2 into the rule f'(x) = 2x-2, obtaining

$f'(2) = 2(2) - 2 = 2$.

(c) The slope of the line tangent to the curve at the point where $x = 2$ is $f'(2) = 2$.

(d) Since $f(x) = x^2 - 2x + 4$, $f(2) = (2)^2 - 2(2) + 4 = 4$. Thus the point (2,4) lies on the curve and is the point of tangency. The slope of the tangent line at this point is given by the derivative at $x = 2$, i.e. $f'(2) = 2$. Using the point-slope form of a line, we have

$$y - 4 = 2(x - 2)$$
$$y - 4 = 2x - 4$$
or $y = 2x$ as an equation for the tangent line.

(e) The graph of f is the graph of the parabola $y = x^2 - 2x + 4$. The turning point of the parabola $y = ax^2 + bx + c$ occurs at $x = -b/2a$. In this case $a = 1$ and $b = -2$, so that the turning point occurs at $x = -(-2)/2(1) = 1$. To obtain the graph we select values of x arbitrarily, but centered around $x = 1$ and determine the corresponding values of y. We make a table

x	-1	0	1	2	3
$y = x^2 - 2x + 4$	7	4	3	4	7

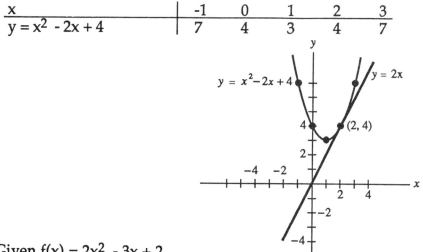

13. Given $f(x) = 2x^2 - 3x + 2$

We will use the alternate definition of the derivative

$$f'(x_0) = \lim_{x \to x_0} \frac{f(x) - f(x_0)}{x - x_0}.$$

For $x_0 = -2$, we first note that $f(x_0) = f(-2) = 2(-2)^2 - 3(-2) + 2 = 16$. Thus

$$f'(-2) = \lim_{x \to -2} \frac{f(x) - f(-2)}{x-(-2)} = \lim_{x \to -2} \frac{(2x^2 - 3x + 2) - 16}{x + 2} =$$

$$\lim_{x \to -2} \frac{2x^2 - 3x - 14}{x+2} = \lim_{x \to -2} \frac{(2x-7)(x+2)}{x+2} = \lim_{x \to -2} (2x - 7) = -11$$

Thus $f'(-2) = -11$

For $x_0 = 0$, we first note that $f(x_0) = f(0) = 2 \cdot 0 - 3 \cdot 0 + 2 = 2$. Thus

$$f'(0) = \lim_{x \to 0} \frac{f(x) - f(0)}{x-0} = \lim_{x \to 0} \frac{(2x^2 - 3x + 2) - 2}{x} = \lim_{x \to 0} \frac{2x^2 - 3x}{x}$$

$$= \lim_{x \to 0} (2x - 3) = -3$$

Thus $f'(0) = -3$

For $x_0 = 3$, we first note that $f(x_0) = f(3) = 2 \cdot 3 - 3 \cdot 3 + 2 = 11$. Thus

$$f'(3) = \lim_{x \to 3} \frac{f(x) - f(3)}{x - 3} = \lim_{x \to 3} \frac{(2x^2 - 3x + 2) - 11}{x - 3} = \lim_{x \to 3} \frac{2x^2 - 3x - 9}{x - 3}$$

$$= \lim_{x \to 3} \frac{(2x +3)(x - 3)}{x - 3} = \lim_{x \to 3} (2x + 3) = 9$$

Thus $f'(3) = 9$

17. Given $f(x) = 2x^2 - 3x + 2$

    The function is a quadratic function fo the form $f(x) = ax^2 + bx + c$ where $a = 2$ and $b = -3$. Equation (6) tells us the derivative is given by $f'(x) = 2ax + b$. Hence $f'(x) = 2(2)x + (-3) = 4x - 3$.

(a) If the slope is 5, then the derivative is 5. Thus we are asked to solve

$$4x-3 = 5$$
$$4x = 8$$
$$x = 2$$

To find the y-coordinate, we substitute x = 2 back into the original function y = f(2) = 2(2)^2 - 3(2) + 2 = 4. Thus, at the point (2,4), the function f(x) = 2x^2 - 3x + 2 will have a tangent line that has a slope of 5.

(b) If the tangent line is horizontal, then its slope is zero. Thus the derivative is zero. We solve for x, knowing

$$f'(x) = 4x-3 = 0$$
$$x = 3/4$$

To find the y-coordinate, we substitute x = 3/4 back into the original function y = f(3/4) = 2(3/4)^2 - 3(3/4) + 2 = 7/8. Thus, at the point (3/4, 7/8), the function f(x) = 2x^2 - 3x+ 2 will have a horizontal tangent line.

(c) The line 3x + y - 5 = 0 is written in slope-intercept form as y = -3x + 5. Thus its slope is -3. Any line parallel to this line must also have a slope of -3. We are asked to find the point at which the tangent line to the curve f(x) = 2x^2 - 3x+ 2 has a slope of -3.

Since f'(x) = 4x-3, we must solve for x, knowing

$$4x-3 = -3$$
$$4x = 0$$
$$x = 0$$

To find the y-coordinate we substitute back into the original y = f(0) = 2(0)^2 - 3(0) + 2 = 2. Thus, at the point (0,2) the tangent line to the curve f(x) = 2x^2 - 3x+ 2 will have a slope of -3 and hence be parallel to the line 3x + y - 5 = 0.

21. Use the definition $N'(x) = f'(x_0) = \lim_{h \to 0} \frac{N(x+h) - N(x)}{h}$ to show that

N'(x) = 40-2x.

(a) We substitute x = 10 obtaining N'(10) = 40 -2(10) = 20. Thus, 10 days after its start, the number of cases increases at a rate of 20 cases per day.

(b) We substitute x = 20 obtaining N'(2)) = 40-2(20) = 0. Thus, 20 days after its start, the epidemic displays a rate of change of 0 cases per day.

(c) We substitute x = 30 obtaining N'(30) = 40-2(30) = -20. Thus, 30 days after its start, the number of cases decreases at a rate of 20 cases per day.

23. The graph of f is the graph of

$$y = \begin{cases} -3x + 7 & \text{if } x \geq 1 \\ 4x & \text{if } x < 1 \end{cases}$$

We sketch the graph in two stages. First, we sketch the linear graph y = 4x for x < 1. Next, we sketch the linear graph y = -3x + 7 for x ≥ 1. The resulting sketch is

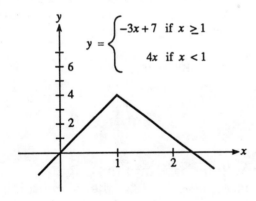

Since the graph of f has a sharp corner at x = 1, f is not differentiable at x = 1.

Exercise Set 3.2, (Page 147)

3. Given $f(x) = \sqrt{x-1}$, $x_0 = 10$, $x_1 = 17$

$$\Delta x = x_1 - x_0 = 17 - 10 = 7$$
$$\Delta y = f(x_1) - (x_0) = f(17) - f(10) = \sqrt{17-1} - \sqrt{10-1} = \sqrt{16} - \sqrt{9}$$
$$= 4 - 3 = 1$$

13. Given $f(x) = 3x^4 - 2x + 2$, we find the derivative.

$$\frac{d}{dx}[3x^4 - 2x + 2] = \frac{d}{dx}(3x^4) - \frac{d}{dx}(2x) + \frac{d}{dx}(2)$$
$$= 3\frac{d}{dx}(x^4) - 2\frac{d}{dx}(x) + \frac{d}{dx}(2)$$
$$= 3 \cdot 4x^3 - 2(1) + 0 = 12x^3 - 2$$

17. Given $g(u) = 2u - 5$, we find the derivative.

$$\frac{d}{du}(2u - 5) = \frac{d}{du}(2u) - \frac{d}{du}(5)$$
$$= 2\frac{d}{du}(u) - \frac{d}{du}(5)$$
$$= 2(1) - 0 = 2$$

23. Given $h(x) = (2x + 3)(x^2 - 1)$

We multiply the factors to get h(x) in polynomial form.
$h(x) = 2x^3 + 3x^2 - 2x - 3$

$$\frac{d}{dx}[2x^3 + 3x^2 - 2x - 3] = \frac{d}{dx}(2x^3) + \frac{d}{dx}(3x^2) - \frac{d}{dx}(2x) - \frac{d}{dx}(3)$$
$$= 2\frac{d}{dx}(x^3) + 3\frac{d}{dx}(x^2) - 2\frac{d}{dx}(x) - \frac{d}{dx}(3)$$
$$= 2(3x^2) + 3(2x) - 2(1) - 0$$
$$= 6x^2 + 6x - 2$$

25. $f(x) = 2x^3 - 3x^2 + 2$. The derivative is

$$f'(x) = 2(3x^2) - 3(2x) = 6x^2 - 6x$$

(a) $f'(-2) = 6(-2)^2 - 6(-2) = 24 + 12 = 36$
(b) $f'(0) = 6(0)^2 - 6(0) = 0$
(c) $f'(3) = 6(3)^2 - 6(3) = 54 - 18 = 36$

29. Given $f(x) = x^3 + x^2 - 8x + 2$. The derivative is

$$f'(x) = 3x^2 + 2x - 8$$

(a) If $f'(x) = -8$, we have

$$3x^2 + 2x - 8 = -8$$
$$3x^2 + 2x = 0$$
$$x(3x + 2) = 0$$
$$x = 0 \text{ or } x = -2/3$$

(b) If f'(x) = 0, we have

$$3x^2 + 2x - 8 = 0$$
$$(3x - 4)(x + 2) = 0$$
$$x = 4/3 \text{ or } x = -2$$

(c) If f'(x) = 8, we have

$$3x^2 + 2x - 8 = 8$$
$$3x^2 + 2x - 16 = 0$$
$$(3x + 8)(x-2) = 0$$
$$x = -8/3 \text{ or } x = 2$$

33. The derivative of A with respect to r gives the rate of change of the area with respect to the radius. Since $A = \pi r^2$, then $A' = \pi (2r) = 2\pi r$. (Remember that π is a constant.)

Exercise Set 3.3, (Page 156)

3. Given $f(x) = x^4 (3x^2 + 7)$. Using the product rule we have

$$f'(x) = x^4 \frac{d}{dx}(3x^2 + 7) + (3x^2 + 7)\frac{d}{dx}(x^4)$$
$$= x^4(6x) + (3x^2 + 7)(4x^3) = 6x^5 + 12x^5 + 28x^3 = 18x^5 + 28x^3$$

If we first multiply out the factors we have

$$f(x) = x^4(3x^2 + 7) = 3x^6 + 7x^4$$

Using the rules for polynomial functions, we have

$$f'(x) = 3(6x^5) + 7(4x^3) = 18x^5 + 28x^3$$

7. Given $g(s) = (s^3 - 5)(s^2 + s + 3)$. Using the product rule, we have

$$g'(s) = (s^3 - 5)\frac{d}{ds}(s^2 + s + 3) + (s^2 + s + 3)\frac{d}{ds}(s^3 - 5)$$
$$= (s^3 - 5)(2s + 1) + (s^2 + s + 3)(3s^2)$$

$$= 2s^4 + s^3 - 10s - 5 + 3s^4 + 3s^3 + 9s^2$$
$$= 5s^4 + 4s^3 + 9s^2 - 10s - 5$$

11. $h(u) = \frac{u^2}{u^2 + 6}$. Using the quotient rule, we have

$$h'(u) = \frac{(u^2 + 6)\frac{d}{du}u^2 - u^2\frac{d}{du}(u^2 + 6)}{(u^2 + 6)^2}$$

$$= \frac{(u^2 + 6)(2u) - u^2(2u)}{(u^2 + 6)^2} = \frac{2u^3 + 12u - 2u^3}{(u^2 + 6)^2} = \frac{12u}{(u^2 + 6)^2}$$

17. We rewrite f using negative exponents as

$$f(x) = 2x^{-2} - 3x^{-4} + 5x^4 + 7$$

Then

$$f'(x) = 2(-2)x^{-3} - 3(-4)x^{-5} + 5(4)x^3 + 0$$
$$= 4x^{-3} - 12x^{-5} + 20x^3 = \frac{-4}{x^3} + \frac{12}{x^5} + 20x^3$$

23. We rewrite h using fractional exponents as

$$h(t) = (12t^3 - 4t)(t^{2/3} + 5t)$$

We next apply the product rule, obtaining

$$h'(t) = (12t^3 - 4t)\frac{d}{dt}[(t^{2/3} + 5t)] + (t^{2/3} + 5t)\frac{d}{dt}[(12t^3 - 4t)]$$
$$= (12t^3 - 4t)[\frac{2}{3}t^{-1/3} + 5] + (t^{2/3} + 5t)(12t^2 - 4)$$
$$= (8t^{8/3} - \frac{8}{3}t^{2/3} + 60t^3 - 20t) + (36t^{8/3} + 180t^3 - 4t^{2/3} - 20t)$$
$$= 44t^{8/3} - \frac{20}{3}t^{2/3} + 240t^3 - 40t$$

Alternately, multiplying factors, we can write h as

$$h(t) = 12t^{11/3} - 4t^{5/3} + 60t^4 - 20t^2$$

And differentiation yields

$$h'(t) = 12\left(\frac{11}{3}\right)t^{8/3} - 4\left(\frac{5}{3}\right)t^{2/3} + 60(4)t^3 - 20(2)t$$

60 Study Guide

$$= 44t^{8/3} - \frac{20}{3}t^{2/3} + 240t^3 - 40t$$

25. We first rewrite h using fractional exponents as

$$h(x) = \frac{3x^{1/2} + x^2}{2 + 4x^{1/2}}$$

Using the quotient rule, we obtain

$$h'(x) = \frac{(2 + 4x^{1/2})\frac{d}{dx}(3x^{1/2} + x^2) - (3x^{1/2} + x^2)\frac{d}{dx}(2 + 4x^{1/2})}{(2 + 4x^{1/2})^2}$$

$$= \frac{(2 + 4x^{1/2})(\frac{3}{2}x^{-1/2} + 2x) - (3x^{1/2} + x^2)[4\frac{1}{2}x^{-1/2}]}{(2 + 4\sqrt{x})^2}$$

Multiply the numerator and denominator by $\sqrt{x} = x^{1/2}$, obtaining

$$h'(x) = \frac{(2 + 4x^{1/2})(\frac{3}{2} + 2x^{3/2}) - (3x^{1/2} + x^2)2}{\sqrt{x}(2 + 4\sqrt{x})^2}$$

$$= \frac{(3 + 6x^{1/2} + 4x^{3/2} + 8x^2) - (6x^{1/2} + 2x^2)}{\sqrt{x}(2 + 4\sqrt{x})^2}$$

$$= \frac{3 + 4x^{3/2} + 6x^2}{\sqrt{x}(2 + 4\sqrt{x})^2}$$

31. Given $h(x) = \frac{2x-4}{3x+2}$. Using the quotient rule, we have

$$h'(x) = \frac{(3x+2)\frac{d}{dx}(2x-4) - (2x-4)\frac{d}{dx}(3x+2)}{(3x+2)^2}$$

$$= \frac{(3x+2)(2) = (2x-4)(3)}{(3x+2)^2}$$

$$= \frac{6x + 4 - 6x + 12}{(3x+2)^2} = \frac{16}{(3x+2)^2}$$

(a) $h'(3) = \dfrac{16}{[3(-3) + 2]^2} = \dfrac{16}{(-7)^2} = \dfrac{16}{49}$

(b) $h'(0) = \dfrac{16}{[3(0) + 2]^2} = \dfrac{16}{4} = 4$

(c) $h'(2) = \dfrac{16}{[3(2) + 2]^2} = \dfrac{16}{64} = \dfrac{1}{4}$

33. Given $f(x) = \dfrac{x}{x-1}$. The slope of the tangent line is given by the derivative. Using the quotient rule, we have

$$f'(x) = \dfrac{(x-1)\dfrac{d}{dx}(x) - (x)\dfrac{d}{dx}(x-1)}{(x-1)^2}$$

$$= \dfrac{(x-1)(1) - (x)(1)}{(x-1)^2}$$

$$= \dfrac{x - 1 - x}{(x-1)^2} = \dfrac{-1}{(x-1)^2}$$

(a) At $x = 0$

$$f'(0) = \dfrac{-1}{(0-1)^2} = -1$$

(b) At $x = -1$

$$f'(-1) = \dfrac{-1}{(-1-1)^2} = \dfrac{-1}{4}$$

(c) At $x = 3$

$$f'(3) = \dfrac{-1}{(3-1)^2} = \dfrac{-1}{4}$$

Exercise Set 3.4, (Page 164)

1. (a) $MC = C'(x) = 40 - 2x/20 = 40 - x/10$
   $MR = R'(x) = 80 + 2x/10 = 80 + x/5$

   (b) Since $C'(30) = 40 - 30/10 = 37$, the cost is increasing at a rate of \$37 per motorcycle when the level of production is 30 motorcycles.

(c) Since R'(30) = 80 + 30/5 = 86, the revenue is increasing at a rate of $86 per motorcycle when the level of production is 30 motorcycles.

(d) The cost of manufacturing the 31st motorcycle is approximately C'(30) = $37.

(e) The revenue from selling the 31st motorcycle is approximately R'(30) = $86.

3. (a) The profit function P(x) = R(x) - C(x). Thus

$$P(x) = (80x + \frac{x^2}{10}) - (2000 + 40x - \frac{x^2}{20})$$

$$P(x) = 40x + \frac{3x^2}{20} - 2000$$

The marginal profit is P'(x). Thus

$$P'(x) = 40 + \frac{6}{20}x = 40 + \frac{3}{10}x$$

(b) Since $P'(40) = 40 + \frac{3}{10}(40) = 40 + 12 = 52$, the rate of change in profit relative to a one-unit increase in production is $52 per motorcycle when the production level is 40 motorcycles.

(c) The profit from manufacturing and selling the 41st motorcycle is approximatley P'(40) = $52.

7. Given $C(x) = 50 + 3x + 2x^{1/2}$ where C(x) is the cost in dollars of manufacturing x cameras per month. We want to find x when the marginal cost, C'(x), is $3.25 per camera. Thus

$$C'(x) = 3 + 2(1/2)^{-1/2} = 3.25$$

or

$$3 + \frac{1}{\sqrt{x}} = 3.25$$

$$\frac{1}{\sqrt{x}} = .25$$

Squaring both sides gives

$$\frac{1}{x} = .0625$$

or $x = \frac{1}{.0625} = 16$

The marginal cost will be $3.25, when 16 cameras per month are manufactured.

Exercise Set 3.5, (Page 172)

1. Let $u = 3x + 2$; then $y = u^6$

5. Using formula (6).

$$\frac{dy}{dx} = 20(3x^2 + 1)^{19} \frac{d}{dx}(3x^2 + 1)$$

$$= 20(3x^2 + 1)^{19}(6x) = 120x(3x^2 + 1)^{19}$$

13. We rewrite y as $x^2(4x - 2x^2)^{1/2}$ and use the product rule and then formula (6) to obtain

$$\frac{dy}{dx} = x^2 \frac{d}{dx}[(4x - 2x^2)^{1/2}] + (4x - 2x^2)^{1/2} 2x$$

$$= x^2 \frac{1}{2}[(4x - 2x^2)^{-1/2}] \frac{d}{dx}(4x - 2x^2) + (4x - 2x^2)^{1/2} 2x$$

$$= x^2 \frac{1}{2}(4x - 2x^2)^{-1/2}(4-4x) + \sqrt{4x - 2x^2}\, 2x$$

$$= \frac{2x^2 - 2x^3}{\sqrt{4x - 2x^2}} + \frac{(4x - 2x^2) 2x}{\sqrt{4x - 2x^2}}$$

$$= \frac{10x^2 - 6x^3}{\sqrt{4x - 2x^2}} = \frac{x^2(10-6x)}{\sqrt{4x - 2x^2}}$$

17. Using formula (6), we have

$$\frac{dy}{dx} = \frac{1}{5}(5x^3 + 4x^2 - 2x + 4)^{-4/5} \frac{d}{dx}(5x^3 + 4x^2 - 2x + 4)$$

$$= \frac{1}{5}(5x^3 + 4x^2 - 2x + 4)^{-4/5}(15x^2 + 8x - 2)$$

$$= \frac{(15x^2 + 8x - 2)}{5(5x^3 + 4x^2 - 2x + 4)^{4/5}}$$

25. We rewrite y as $4(x^2 + 2x + 1)^{-1/2}$. Using formula (6), we have

$$\frac{dy}{dx} = 4\left(\frac{-1}{2}\right)(x^2 + 2x + 1)^{-3/2} \frac{d}{dx}(x^2 + 2x + 1)$$

$$= \frac{-2(2x+2)}{(x^2 + 2x + 1)^{3/2}} = \frac{-4(x+1)}{(x^2 + 2x + 1)^{3/2}}$$

The slope at $x = 2$ is

$$\frac{-4(2+1)}{((2)^2 + 2(2) + 1)^{3/2}} = \frac{-12}{9^{3/2}} = \frac{-12}{27} = \frac{-4}{9}$$

The equation of the tangent line to the graph of the function $y = \frac{4}{\sqrt{x^2 + 2x + 1}}$ at the point $(2, \frac{4}{3})$ is

$$y - \frac{4}{3} = \frac{-4}{9}(x - 2)$$

or

$$y = \frac{-4}{9}x + \frac{20}{9}$$

27. Given $N(t) = \sqrt{2t^3 - 9}$ where N(t) is the number of radios assembled daily by a typical worker after t hours of training. The rate of learning is given by N'(t). Using formula (6) after rewriting N(t) as $(2t^3 - 9)^{1/2}$, we have

$$N'(t) = \frac{1}{2}((2t^3 - 9)^{-1/2} \frac{d}{dt}(2t^3 - 9)$$

$$= \frac{6t^2}{2\sqrt{2t^3 - 9}} = \frac{3t^2}{\sqrt{2t^3 - 9}}$$

At $t = 2$, $N'(2) = \frac{3(2)^2}{\sqrt{2(2)^3 - 9}} = \frac{12}{\sqrt{25}} = \frac{12}{5} = 2.4$

After 2 hours of training, 2.4 radios per hour are assembled daily by a typical worker.

Exercise Set 3.6, (Page 178)

1. We are given that dx/dt = -3 tons per week and that $y = 200 + 300x^{-1}$. Thus, dy/dx is $-300x^{-2}$. When the weekly supply x is 30 tons, $dy/dx = -300(30)^{-2} = -1/3$ dollars per ton. The chain rule gives the rate of change of price per week as

$$\frac{dy}{dt} = \frac{dy}{dx}\frac{dx}{dt} = \frac{-1}{3}(-3) = 1$$

   Thus, the price per ton y is increasing at a rate of $1 per week.

7. We are given that dv/dt = 12 cubic inches per second.

   (a) We want to find dr/dt when the diameter is 6 inches. Since the diameter is twice the radius, we want to find dr/dt when r = 3.

   The volume of a sphere of radius r is $V = 4/3 \pi r^3$. We see that $dv/dr = 4/3 \pi (3r)^2 = 4\pi r^2$. When r = 3, $dv/dr = 4\pi(3)^2 = 36\pi$.

   The chain rule gives

$$\frac{dv}{dt} = \frac{dv}{dr}\frac{dr}{dt}. \text{ Thus } 12 = 36\pi \frac{dr}{dt} \text{ or}$$

$$\frac{dr}{dt} = \frac{12}{36\pi} = \frac{1}{3\pi} \text{ inches per second.}$$

   (b) If $V = 9\pi/2$ cubic inches, then the formula $V = 4/3 \pi r^3$ gives

$$\frac{9\pi}{2} = \frac{4}{3}\pi r^3$$

$$\frac{27}{8} = r^3$$

$$\frac{3}{2} = r$$

   Thus at r = 3/2, $dv/dr = 4/3 \pi (3r)^2 = 4\pi r^2 = 4\pi(3/2)^2 = 9\pi$

   The chain rule gives

$$\frac{dv}{dt} = \frac{dv}{dr}\frac{dr}{dt}. \text{ Thus}$$

66 Study Guide

$$12 = 9\pi \frac{dr}{dt} \text{ or } \frac{dr}{dt} = \frac{12}{9\pi} = \frac{4}{3\pi} \text{ inches per second.}$$

11. Given the demand equation

    $$p = 800 - 15x + x^2/10$$

    We want to find the instantaneous rate of change in the demand, dx/dt, when $dp/dt = -15$ and $x = 60$. The chain rule gives

    $$\frac{dp}{dt} = \frac{dp}{dx}\frac{dx}{dt}$$

    Now $\frac{dp}{dx} = -15 + \frac{2x}{10} = -15 + \frac{x}{5}$. At $x = 60$ $\frac{dp}{dx} = -15 + \frac{60}{5} = -15 + 12 = -3$.
    Thus, substitution in the chain rule yields

    $$-15 = -3\frac{dx}{dt} \text{ or } \frac{dx}{dt} = 5$$

    The demand is increasing at the rate of 5 units per month.

Exercise Set 3.7, (Page 183)

5. We must differentiate both sides of the given equation with respect to x and treat y as an unknown differentiable function of x. Since $x^2 y^2$ is a product, we have

    $$\frac{d}{dx}[x^2 y^2] = x^2 \frac{d}{dx}[y^2] + y^2 \frac{d}{dx}[x^2]$$

    $$= x^2 \, 2y \, y' + y^2 \, 2x$$

    Also,

    $$\frac{d}{dx}[3y^3] = 3(3y^2 \, y') = 9y^2 \, y'$$

    Thus, we obtain

    $$2x^2 \, y \, y' + 2xy^2 - 4x + 9y^2 \, y' = 0$$

    We can now solve for y':

$$(2x^2y + 9y^2)y' = 4x - 2xy^2$$

Hence

$$y' = \frac{dy}{dx} = \frac{4x-2xy^2}{2x^2y + 9y^2}$$

7. We must differentiate both sides of the given equation with respect to x and treat y as an unknown differentiable function of x. Note that each term on the left hand side requires the chain rule.

Specifically,

$$\frac{d}{dx}[(x+y)^2] = 2(x+y)\frac{d}{dx}[x+y]$$

$$= 2(x+y)(1+y')$$

and

$$\frac{d}{dx}[(x-y)^2] = 2(x-y)\frac{d}{dx}[x-y]$$

$$= 2(x-y)(1-y')$$

Thus, we obtain

$$2(x+y)(1+y') + 2(x-y)(1-y') = 0$$

or after dividing by 2 and expanding

$$(x+y) + (x+y)y' + (x-y) - (x-y)y' = 0$$

$$2x + 2y\,y' = 0$$

Solving for y' yields

$$y' = \frac{dy}{dx} = \frac{-x}{y}$$

15. We must differentiate both sides of the given equation with respect to x and treat y as an unknown differentiable function of x. Since xy^2 is a product, we have

$$\frac{d}{dx}[xy^2] = x\frac{d}{dx}[y^2] + y^2\frac{d}{dx}[x]$$

$$= x \cdot 2y \, y' + y^2$$

Then, we obtain

$$3y^2 y' - 2xy \, y' - y^2 + 4 = 0$$

We now solve for y' as

$$y'(3y^2 - 2xy) = y^2 - 4$$

or

$$y' = \frac{y^2 - 4}{3y^2 - 2xy}$$

Hence

$$\left.\frac{dy}{dx}\right|_{(1,-1)} = \frac{(-1)^2 - 4}{3(-1)^2 - 2(1)(-1)} = \frac{-3}{5}$$

19. Since the price p (in dollars per pound and the number x of pounds demanded per week are unknown functions of the time t (in weeks) related by px + 16 p - 800 = 0, we know that their rates are related by

$$\frac{d}{dt}[px] + 16\frac{dp}{dt} = 0$$

Thus,

$$p\frac{dx}{dt} + x\frac{dp}{dt} + 16\frac{dp}{dt} = 0$$

We are given p = 5 and dp/dt = 0.06. Note from the original equation that when p = 5

$$5x + (16)(5) - 800 = 0$$
$$5x = 800 - 80 = 720$$
$$x = 144$$

so that from the differentiated equation we have

$$5\frac{dx}{dt} + 144(0.06) + 16(0.06) = 0$$

3: The Derivative 69

$$5\frac{dx}{dt} = -9.6$$

$$\frac{dx}{dt} = -1.92$$

Thus, the demand is decreasing by 1.92 pounds per week.

23. 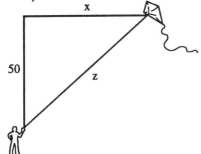 From the Pythagorean theorem, we have $x^2 + (50)^2 = z^2$ (1)

We are told that $dx/dt = 4$. We want to find dz/dt, the rate at which the boy is paying out the string, when $z = 130$.

Differentiating equation (1) with respect to t, gives

$$\frac{d}{dt}(x^2) + \frac{d}{dt}(50^2) = \frac{d}{dt}(z^2)$$

$$2x\frac{dx}{dt} + 0 = 2z\frac{dz}{dt}$$

Thus, $\frac{dz}{dt} = \frac{x}{z}\frac{dx}{dt}$

When $z = 130$, we have

$$x^2 + (50)^2 = (130)^2$$
$$x^2 + 2500 = 16900$$
$$x^2 = 14400$$
$$x = 120$$

Substituting into (2) gives $\frac{dz}{dt} = \frac{120}{130}(4) = \frac{48}{13} = 3.69$

Thus the boy is paying out the string at the rate of 3.69 feet per second.

Review Exercises, (Page 186)

3. Given $f(x) = 2 - 3x$. Using the definition of the derivative, we have

$$f(x + h) = 2 - 3x - 3h$$

$$f(x+h) - f(x) = 2 - 3x - 3h - (2 - 3x) = -3h$$

$$\frac{f(x+h) - f(x)}{h} = \frac{-3h}{h} = -3$$

$$f'(x) = \lim_{h \to 0} \frac{f(x+h) - f(x)}{h} = \lim_{h \to 0} (-3) = -3$$

9. We first rewrite $f(x) = 2\sqrt{x} - \frac{1}{x}$ using negative and fractional exponents as

$$f(x) = 2x^{1/2} - x^{-1}$$

From this expression we obtain

$$f'(x) = 2[\tfrac{1}{2}x^{-1/2}] - (-x^{-2}) = x^{-1/2} + x^{-2} = \frac{1}{\sqrt{x}} + \frac{1}{x^2}$$

13. Using the quotient rule, we obtain

$$\frac{dy}{dx} = \frac{(x^2+1)\frac{d}{dx}[x-1] - (x-1)\frac{d}{dx}[x^2+1]}{[x^2+1]^2}$$

$$= \frac{(x^2+1)(1) - (x-1)(2x)}{[x^2+1]^2} = \frac{-x^2 + 2x + 1}{[x^2+1]^2}$$

17. The tangent line to the graph of f is horizontal when the derivative at the point of tangency is zero. Since $f(x) = 2x^3 + 5x^2 - 4x$, $f'(x) = 6x^2 + 10x - 4$. We factor the derivative and set it equal to zero, obtaining

$$f'(x) = (3x-1)(2x+4) = 0$$

Thus, $3x - 1 = 0$ or $2x + 4 = 0$. The tangent lines are horizontal at $x = 1/3$ and $x = -2$. The points of tangency are $(1/3, -19/27)$ and $(-2, 12)$.

19. Using the formula for the derivative of a function to a power, we obtain

$$\frac{dy}{dx} = -8(2x+3)^{-9} \frac{d}{dx}[2x+3]$$
$$= -8(2x+3)^{-9}(2) = -16(2x+3)^{-9}$$

21. We first rewrite $y = \sqrt{x^2 + 3x}$ using fractional exponents as $y = (x^2 + 3x)^{1/2}$. Next, use the formula for the derivative of a function to a power, obtaining

$$\frac{dy}{dx} = \frac{1}{2}(x^2+3x)^{-1/2}\frac{d}{dx}[x^2+3x]$$

$$= \frac{1}{2} \cdot \frac{1}{\sqrt{x^2+3x}}(2x+3) = \frac{2x+3}{2\sqrt{x^2+3x}}$$

27. (a) The instantaneous velocity is the rate of change of distance with respect to time, that is, ds/dt. Since $s = 2t^4 - 80t^3$, $ds/dt = 8t^3 - 240t^2$

    (b) The object is set at rest when its instantaneous velocity is zero. Thus, we set

    $$\frac{ds}{dt} = 8t^3 - 240t^2 = 0$$

    so that $8t^2(t - 30) = 0$. The object is at rest initially ($t = 0$) and comes to rest again after 30 seconds.

29. (a) The marginal cost is $C'(x) = 0.1x + 100$. When $x = 2000$, the marginal cost is $C'(2000) = 0.1(2000) + 100 = 300$ cents, or \$3.

    (b) The cost of making the 2001 pen is approximately $C'(2000) = \$3$.

35. We must differentiate both sides of the given equation with respect to x and treat y as an unknown differentiable function of x.

    $$2x + 2\frac{d}{dx}(xy) - \frac{d}{dx}(y^2) = 0$$

    $$2x + 2[x\frac{d}{dx}(y) + y\frac{d}{dx}(x)] - 2y\,y' = 0$$

    $$2x + 2x\,y' + 2y - 2y\,y' = 0$$

    Dividing by 2 and solving for y' yields

    $$(x - y)\,y' = -(x + y)$$

    or

72 *Study Guide*

$$y' = \frac{-(x+y)}{x-y} = \frac{x+y}{y-x}$$

Chapter Test, (Page 189)

1. Given $f(x) = x - 2x^2$. To find $f'(x)$ by the four-step procedure we first find $f(x + h)$ by replacing every occurrence of x in the rule for $f(x)$ by $x + h$. Thus

    Step 1. $f(x + h) = (x + h) - 2(x + h)^2 = x + h - 2(x^2 + 2xh + h^2)$
    $= x + h - 2x^2 - 4xh - 2h^2$

    Step 2. $f(x + h) - f(x) = (x + h - 2x^2 - 4xh - 2h^2) - (x - 2x^2)$
    $= h - 4xh - 2h^2$

    Step 3. $\frac{f(x + h) - f(x)}{h} = \frac{h - 4xh - 2h^2}{h} = 1 - 4x - 2h$

    Step 4. $\lim_{h \to 0} \frac{f(x + h) - f(x)}{h} = \lim_{h \to 0} (1 - 4x - 2h) = 1 - 4x$

    So $f'(x) = 1 - 4x$. To find $f'(1)$, we substitute $x = 1$ into the rule $f'(x) = 1 - 4x$, obtaining $f'(1) = 1 - 4(1) = -3$.

3. The function $f(x) = |x + 3|$ is not differentiable at $x = -3$ since it has a sharp corner there. The function is continuous everywhere. *i.e.* for all real values of x.

5. Given $f(x) = 49^3 + 2x$, we find the derivative.

    $$\frac{d}{dx}[49^3 + 2x] = \frac{d}{dx}[49^3] + \frac{d}{dx}[2x] = 0 + 2 = 2$$

7. Given $g(x) = \frac{3x + 8}{x^2 - 1}$. Using the quotient rule, we have

    $$g'(x) = \frac{(x^2 - 1)\frac{d}{dx}(3x + 8) - (3x + 8)\frac{d}{dx}(x^2 - 1)}{(x^2 - 1)^2}$$

$$= \frac{(x^2-1)(3) - (3x+8)(2x)}{(x^2-1)^2} = \frac{3x^2 - 3 - 6x^2 - 16x}{(x^2-1)^2}$$

$$= \frac{(-3x^2 + 16x + 3)}{(x^2-1)^2}$$

9. Given $f(x) = (3x^4 + 7x + 2)^8 (x^3 - 4x^2)$. Using the product rule and formula (6), we have

$$f'(x) = (3x^4 + 7x + 2)^8 \frac{d}{dx}(x^3 - 4x^2) + (x^3 - 4x^2) \frac{d}{dx}(3x^4 + 7x + 2)^8$$

$$= (3x^4 + 7x + 2)^8 (3x^2 - 8x) + (x^3 - 4x^2)(8)(3x^4 + 7x + 2)^7 (12x^3 + 7)$$

$$= [3x^4 + 7x + 2]^7 [(3x^4 + 7x + 2)(3x^2 - 8x) + 8(x^3 - 4x^2)(12x^3 + 7)]$$

11. We must differentiate both sides of the given equation with respect to x and treat y as an unknown differentiable function of x. Since $x^2 y^3$ is a product, we have

$$\frac{d}{dx}[x^2 y^3] = x^2 \frac{d}{dx}[y^3] + y^3 \frac{d}{dx}[x^2]$$

$$= x^2 (3y^2) y' + y^3 (2x)$$

Also

$$\frac{d}{dx}[-3y^2] = -3(2y) y' = -6y \, y'$$

Thus, we obtain

$$3x^2 y^2 y' + 2y^3 x - 6y \, y' = 4x^3$$

$$(3x^2 y^2 - 6y) y' = 4x^3 - 2y^3 x$$

$$y' = \frac{4x^3 - 2y^3 x}{3x^2 y^2 - 6y}$$

At the point (2,1) we obtain the slope of the tangent line by substituting $x = 2$ and $y = 1$ into the derivative. Hence

$$m = y' = \frac{4(2)^3 - 2(1)^3(2)}{3(2)^2(1)^2 - 6(1)} = \frac{32-4}{12-6} = \frac{28}{6} = \frac{14}{3}$$

The equation of the tangent line in point-slope form is

$$y - y_1 = m(x - x_1)$$

Substituting $x_1 = 2$, $y_1 = 1$, and $m = 14/3$, we have

$$y - 1 = \frac{14}{3}(x - 2)$$

$$y - 1 = \frac{14}{3}x - \frac{28}{3}$$

$$y = \frac{14}{3}x - \frac{25}{3}$$

4 Applications of the Derivative

Key Ideas for Review

* If f'(x) > 0 on (a, b), then f is increasing on (a, b); if f'(x) < 0, then f is decreasing on (a, b).

* f has a relative maximum at x = c if there is an open interval containing c such that for any x in this interval, f(c) ≥ f(x); f has a relative minimum at x = c if there is an open interval containing c such that for any x in this interval, f(c) ≤ f(x). The value f(c) is the relative extreme value.

* If f is differentiable at c and f has a relative extremum at c, then f'(c) = 0.

* If f is defined at c and either f'(c) = 0 or f'(c) does not exist, then c is called a critical number of f.

* The critical numbers of f give candidates for the values of x at which f may have relative extrema.

* First derivative test: If the sign of f' changes from plus to minus at c, then f has a relative maximum at c; if the sign of f' changes from minus to plus at c, then f has a relative minimum at c.

* Extreme value theorem: If f is continuous on [a, b], then f has an absolute maximum and an absolute minimum on [a, b].

* To find the absolute extrema of a continuous function f over a closed interval, check the values of f at the critical numbers of f in the interval and at the endpoints of the interval.

* If f"(x) > 0 for all x in (a, b), then the curve y = f(x) is concave upward over (a, b). If f"(x) < 0, then y = f(x) is concave downward over (a, b).

* A point (c, f(c)) on a curve y = f(x) where the curve changes from concave upward to concave downward or vice versa is called an inflection point of the curve.

* Second derivative test: Let c be a critical number of f such that $f'(c) = 0$. If $f''(c) > 0$, then f has a relative minimum at c. If $f''(c) < 0$, then f has a relative maximum at c.

* When the extreme value theorem is not applicable, the following theorem is useful in solving optimization problems in which the function has only one critical number in the interval. Theorem: Suppose f is continuous on the interval I, and I contains only one critical number c of f. If f has a relative maximum (relative minimum) at c, then f(c) is the absolute maximum (absolute minimum) value of f on I.

* The basic procedures for sketching a continuous curve $y = f(x)$ are summarized in Table 1 of Section 4.5.

* If f is differentiable at $x = x_0$ and h is small, then

    $$f(x_0 + h) \approx f(x_0) + f'(x_0)h.$$

* $df = f'(x_0)dx$, and if $y = f(x)$, $dy = f'(x_0)dx$.

* If $y = f(x)$, then

    $\Delta x = dx$ and

    $\Delta f = f(x_0 + dx) - f(x_0) \approx f'(x_0)dx = dy.$

* If a function f is continuous on [a,b] and has opposite signs at a and b, then it must have a zero somewhere in the interval (a,b).

* In the Newton-Raphson algorithm,

    $$x_{n+1} = x_n - \frac{f(x_n)}{f'(x_n)}$$

* The x-intercepts (where $y = 0$) and y-intercepts (where $x = 0$) are often important points of a curve to be sketched. Excluded regions and symmetry are also helpful.

* The points of discontinuity of an otherwise continuous function f separate the graph of f into separate "continuous" pieces.

4: *Applications of the Derivative* 77

*   If $\lim_{x \to \pm\infty} f(x)$ is a finite number, then the graph of f has a horizontal asymptote. If $\lim_{x \to a^+} f(x)$ or $\lim_{x \to a^-} f(x)$ is infinite, then the graph of f has a vertical asymptote at x = a.

Exercise Set 4.1, (Page 196)

5. We find the derivative of f and determine the open intervals where f'(x) is positive and where it is negative. We have

   $$f'(x) = 2x - 4 = 2(x - 2)$$

   Thus, f'(x) > 0 when x > 2 and f'(x) < 0 when x < 2 so f is increasing on (2,∞) and decreasing on (-∞, 2). Note that f'(2) = 0 so that the graph of f has a horizontal tangent at x = 2.

9. We find the derivative of f and determine the open intervals where f'(x) is positive and where it is negative. We have

   $$f'(x) = x^2 - 1 = (x + 1)(x - 1)$$

   Thus, f'(x) > 0 when (x + 1) and (x - 1) have the same sign and f'(x) < 0 when (x + 1) and (x - 1) have opposite signs. A graphical presentation of the sign analysis appears in the figure below

   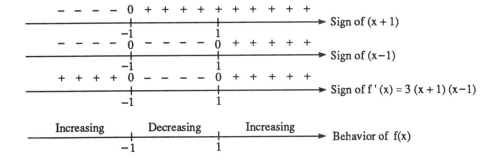

   Thus, f is increasing on (-∞, -1) and on (1, ∞). Also, f is decreasing on (-1, 1). The graph of f has horizontal tangents at x = -1 and at x = 1 since f'(-1) = f'(1) = 0.

13. We find the derivative of f and determine the open intervals where f'(x) is positive and where it is negative. We have

$$f'(x) = \frac{1}{4}(4x^3) - \frac{1}{2}(2x) = x^3 - x = x(x^2 - 1)$$

$$f'(x) = x(x-1)(x+1)$$

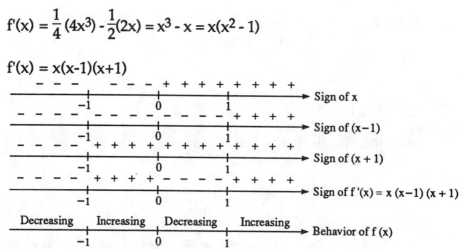

$f'(x) > 0$ when the product of the three factors is positive, and $f'(x) < 0$ when the product of the three factors is negative.

Thus, $f(x)$ is increasing on the interval $(-1,0)$ and on $(1, \infty)$ and $f(x)$ is decreasing on the interval $(-\infty, -1)$ and on $(0,1)$. $f(x)$ has horizontal tangents at $x = 0$, $x = 1$, and $x = -1$ since $f'(0) = f'(1) = f'(-1) = 0$.

19. The graph of $f(x) = x^2 - 4x + 2$ is the graph of $y = x^2 - 4x + 2$. This is a parabola since it is of the form $y = ax^2 + bx + c$ where $a = 1$, $b = -4$, and $c = 2$. Since $a > 0$, the parabola opens upwards. The vertex is the point where the tangent line is horizontal. From exercise 5, it was found to be at $x = 2$. Since $f(2) = -2$, the vertex is at $(2, -2)$. Note that $f(x)$ is decreasing when $x < 2$ and increasing when $x > 2$.

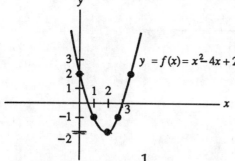

21. The graph of $f(x) = \frac{1}{3}x^3 - x$ is the graph of $y = \frac{1}{3}x^3 - x$. From exercise 9, the graph has horizontal tangents at $x = -1$ and $x = 1$. Since $f(-1) = 2/3$ and $f(1) = -2/3$, we have horizontal tangents at the points $(-1, 2/3)$ and $(1, -2/3)$. We note that the function increases when $x < -1$ and when $x > 1$. The function decreases when $-1 < x < 1$. Using this information and the fact that $f(0) = 0$, we sketch the graph.

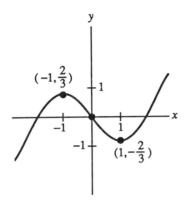

27. The number N(t) of words learned after t weeks of training is given by

$$N(t) = 80t - t^2$$

To find where N(t) is increasing, we must find those values of t for which N'(t) > 0. Since N'(t) = 80 - 2t, we have 80 - 2t > 0 or 80 > 2t or t < 40. To find where N(t) is decreasing, we must find those values of t for which N'(t) < 0. Thus 80 - 2t < 0 or 80 < 2t or t > 40. Thus the number of words learned increases if t < 40 weeks and decreases if t > 40 weeks.

Exercise Set 4.2, (Page 207)

5. We have f'(x) = -2x + 6. To find the critical numbers we set f'(x) = 0 obtaining x = 3. Since f'(x) exists for all values of x, we conclude that this is the only critical number of f. The figure below shows how to determine the sign of f' on either side of the critical number.

Since the sign of f' changes from + to - at x = 3, f has a relative maximum at x = 3.

11. We have f'(x) = 6x² - 6x - 12 = 6(x² - x - 2) = 6(x-2)(x + 1). To find the critical numbers we set f'(x) = 0 obtaining x = 2 and x = -1. Since f'(x) exists for all values of x we conclude that these are the only critical numbers of f. The next figure shows how to determine the sign of f' on either side of the critical numbers.

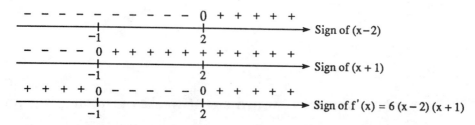

Since the sign of f' changes from + to - at x = -1, f has a relative maximum at x = -1. Similarly, since the sign of f' changes from - to + at x = 2, f has a relative minimum at x = 2.

13. When $f(x) = 1/x^2$, we have $f'(x) = -2/x^3$. Thus f'(x) does not exist for x = 0. However, since f is not defined at x = 0, we conclude that x = 0 is not a critical number of f. Also, since there is no value of x for which f'(x) = 0, we conclude that f has no critical numbers.

17. We have $f'(x) = -3x^2 + 6x - 3 = -3(x^2 - 2x + 1) = -3(x-1)^2$. To find the critical numbers we set f'(x) = 0 obtaining x = 1. Since f'(x) exists for all values of x we conclude that x = 1 is the only critical number of f. The figure below shows how to determine the sign of f' on either side of the critical number.

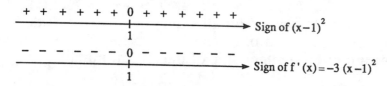

Since the sign of f' does not change at x = 1, there is no extremum at x = 1.

19. Since $f'(x) = \frac{1}{3}x^{-2/3}$, the value x = 0 is the only critical number of f because f'(0) does not exist. If x < 0, then f'(x) > 0 and if x > 0, f'(x) > 0. Since the sign of f' does not change at x = 0, there is no relative extremum at x = 0.

25. We have $f'(x) = 4x^3 - 6x^2 = 2x^2(2x - 3)$. To find the critical numbers of f we set f'(x) = 0 obtaining x = 0 and x = 3/2. Since f'(x) exists for all values of x we conclude that these are the only critical numbers of f. The figure on the next page shows how to determine the sign of f' on either side of the critical numbers.

```
+ + + + 0 + + + + + + + + +
─────────┼─────────────┼──────────→ Sign of x²
         0            3/2
- - - - - - - - - - - 0 + + + + +
─────────┼─────────────┼──────────→ Sign of (2x−3)
         0            3/2
- - - - 0 - - - - - - 0 + + + + +
─────────┼─────────────┼──────────→ Sign of f'(x) = 2x²(2x−3)
         0            3/2
```

Since the sign of f' does not change at x = 0, there is no extremum at x = 0. Also, since the sign of f' changes from - to + at x = 3/2 f has a relative minimum at x = 3/2.

Exercise Set 4.3, (Page 212)

3. When $f(x) = x^2 + 4x - 5$, we have $f'(x) = 2x + 4$. Since $f'(x)$ exists for all values of x, the critical numbers of f occur when $f'(x) = 0$, that is, when x = -2. Since -2 is in the interval [-6,2],, we evaluate f at the critical number and at the endpoints. We have

$$f(-6) = (-6)^2 + 4(-6) - 5 = 7$$
$$f(-2) = (-2)^2 + 4(-2) - 5 = -9$$
$$f(2) = (2)^2 + 4(2) - 5 = 7$$

The absolute maximum of f on [-6,2] is 7 which occurs at x = -6 and at x = 2. The absolute minimum of f on [-6,2] is -9 which occurs at x = -2.

7. When $f(x) = \frac{1}{3}x^3 - x^2 + 10$, we have $f'(x) = x^2 - 2x$. Since $f'(x)$ exists for all values of x, the critical numbers of f occur when $f'(x) = 0$. Setting $x^2 - 2x = 0$ we have $x(x-2) = 0$ or x = 0 and x = 2.

(a) Since only the critical number x = 0 is in the interval [-2,1], we evaluate f only at this critical number and at the endpoints. We have

$$f(-2) = \frac{1}{3}(-2)^3 - (-2)^2 + 10 = \frac{10}{3} = 3\frac{1}{3}$$
$$f(0) = 0 - 0 + 10 = 10$$

$$f(1) = \frac{1}{3}(1)^3 - (1)^2 + 10 = \frac{28}{3} = 9\frac{1}{3}$$

The absolute maximum of f on [-2,1] is 10 which occurs at x = 0. The absolute minimum of f on [-2,1] is $3\frac{1}{3}$ which occurs at x = -2.

(b) Since only the critical number x = 2 is in the interval [1,4], we evaluate f only at this critical number and at the endpoints. We have

$$f(1) = \frac{1}{3}(1)^3 - (1)^2 + 10 = \frac{28}{3} = 9\frac{1}{3}$$

$$f(2) = \frac{1}{3}(2)^3 - (2)^2 + 10 = \frac{26}{3} = 8\frac{2}{3}$$

$$f(4) = \frac{1}{3}(4)^3 - (4)^2 + 10 = \frac{46}{3} = 15\frac{1}{3}$$

The absolute maximum of f on [1,4] is $15\frac{1}{3}$ which occurs at x = 4. The absolute minimum of f on [1,4] is $8\frac{2}{3}$ which occurs at x = 2.

13. When $f(x) = x^{2/3} + 2$, we have $f'(x) = \frac{2}{3}x^{-1/3}$. The value x = 0 is the only critical number of f because f'(0) does not exist. Since 0 is in the interval [-8,1], we evaluate f at the critical number and at the endpoints. We have

$$f(-8) = (-8)^{2/3} + 2 = 4 + 2 = 6$$
$$f(0) = 0 + 2 = 2$$
$$f(1) = (1)^{2/3} + 2 = 3$$

The absolute maximum of f on [-8,1] is 6 which occurs at x = -8. The absolute minimum of f on [-8,1] is 2 which occurs at x = 0.

15. The profit function is $P(x) = 1000x - 25x^2$. We seek the absolute maximum of this function on the interval [0,30]. Since P'(x) = 1000 - 50x, which exists for all values of x, the critical numbers of P occur when P'(x) = 0, that is when x = 20. Since 20 is in the interval [0,30], we evaluate P at the critical number and at the endpoints. We have

$$P(0) = 0$$

4: Applications of the Derivative 83

$$P(20) = 1000(20) - 25(20)^2 = 10{,}000$$
$$P(30) = 1000(30) - 25(30)^2 = 7{,}500$$

The absolute maximum occurs at x = 20, that is, when 20 cars are sold.

Exercise Set 4.4, (Page 221)

3. When $f(x) = x^2 + 3x - 8$, we have

   $$f'(x) = 2x + 3$$

   and

   $$f''(x) = 2$$

   Since f''(x) > 0 for all values of x, the curve is concave upward for all values of x. Thus, since there is no change in concavity, there are no points of inflection.

5. When $f(x) = x^3 + 6x^2 - 15x + 8$, we have

   $$f'(x) = 3x^2 + 12x - 15$$

   and

   $$f''(x) = 6x + 12 = 6(x+2)$$

   We analyze the sign of f" in the figure below

   ```
   - - - - - 0 + + + + +
   ─────────┼───────────→ Sign of (x + 2)
            -2

   - - - - - 0 + + + + +
   ─────────┼───────────→ Sign of f " (x) = 6 (x + 2)
            -2
   ```

   Hence, the curve is concave upward when x > -2 and concave downward when x < -2. Now note that f(-2) = 54 and that the second derivative changes sign at x = -2. Thus, the point (-2, 54) is an inflection point.

9. When $f(x) = x^4 - 4x^3 + 6$, we have

   $$f'(x) = 4x^3 - 12x^2$$

   and

$$f''(x) = 12x^2 - 24x = 12x(x-2)$$

We analyze the sign of f" in the figure below

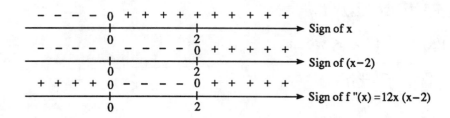

Hence, the curve is concave upward when $x > 2$ and when $x < 0$ and concave downward when $0 < x < 2$. Now note that $f(0) = 6$ and $f(2) = -10$. Since f" changes sign at $x = 0$ and at $x = 2$, the points $(0,6)$ and $(2,-10)$ are inflection points.

11. When $f(x) = x^2 - 6x + 5$, we have

    $$f'(x) = 2x - 6$$

    and

    $$f''(x) = 2$$

    Since f'(x) exists for all values of x, the critical numbers of f occur when $f'(x) = 0$, that is, at $x = 3$. Since $f''(3) = 2 > 0$, f(x) has a relative minimum at $x = 3$. The relative minimum is $f(3) = (3)^2 - 6(3) + 5 = -4$.

13. When $f(x) = 2x^3 - 3x^2 - 12x + 5$, we have

    $$f'(x) = 6x^2 - 6x - 12 = 6(x^2 - x - 2) = 6(x-2)(x+1)$$

    and

    $$f''(x) = 12x - 6$$

    Since f'(x) exists for all values of x, the critical numbers of f occur when $f'(x) = 0$, that is, at $x = 2$, and at $x = -1$. Now

    $$f''(-1) = 12(-1) - 6 = -18 < 0$$

    and

$$f''(2) = 12(2) - 6 = 18 > 0$$

Thus, f has a relative minimum at x = 2 and a relative maximum at x = -1. The relative minimum is f(2) = -15 and the relative maximum is f(-1) = 12.

25. When $f(x) = (x - 3)^{2/3}$, we have

$$f'(x) = \frac{2}{3}(x-3)^{-1/3} = \frac{2}{3(\sqrt[3]{x-3})}$$

and

$$f''(x) = \frac{-2}{9}(x-3)^{-4/3} = \frac{-2}{9(\sqrt[3]{x-3})^4}$$

Since there is no value of x for which f'(x) = 0, the only critical number of f occurs when f' does not exist, that is, when x = 3. We must use the first derivative test at this critical number since the second derivative test does not apply there (why?). We analyze the sign of f' in the figure below.

```
- - - - 0 + + + + +
————————+————————→  Sign of ∛(x-3)
        3

- - - -   + + + + +
————————+————————→  Sign of f'(x) = 2/(3∛(x-3))
        3
```

Since f' changes from - to + at x = 3, f has a relative minimum at x = 3. The relative minimum is f(3) = 0.

29. When $f(x) = x^4/4 - x^2/2 + 2$, we have

$$f'(x) = x^3 - x = x(x^2 - 1) = x(x+1)(x-1)$$

and

$$f''(x) = 3x^2 - 1$$

Since f'(x) exists for all values of x, the only critical numbers of f occur when f'(x) = 0, that is at x = 0, x = 1, and x = -1. We use the second derivative test. Note

$$f''(-1) = 3(-1)^2 - 1 = 2 > 0$$
$$f''(0) = 3(0)^2 - 1 = -1 < 0$$
$$f''(1) = 3(1)^2 - 1 = 2 > 0$$

Thus, f has a relative maximum at $x = 0$ and relative minima at $x = -1$ and $x = 1$. The relative maximum is $f(0) = 2$. The relative minima are $f(1) = f(-1) = 1.75$.

31. The rate of learning $P'(t)$ is given by $P'(t) = 24t - 3t^2$. Since a function decreases when its derivative is negative, the rate of learning decreases when its derivative, $P''(t) = 24 - 6t$ is negative. Note that when

    $$24 - 6t < 0$$

    we have

    $$24 < 6t$$

    or

    $$4 < t$$

    Thus, the rate of learning starts to decrease when $t = 4$.

Exercise Set 4.5, (Page 228)

3. When $f(x) = x^3 - 3x + 3$, we have

    $$f'(x) = 3x^2 - 3 = 3(x^2 - 1) = 3(x + 1)(x - 1)$$

    and

    $$f''(x) = 6x$$

    The critical numbers of f are $x = -1$ and $x = 1$. Using the second derivative test
    $$f''(-1) = 6(-1) = -6 < 0 \text{ and } f''(1) = 6(1) = 6 > 0$$

    so f has a relative maximum at $x = -1$ and a relative minimum at $x = 1$. We analyze the sign of f' in the figure on the next page.

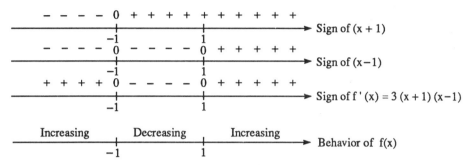

Thus, f is increasing on (-∞, -1) and on (1, ∞) and f is decreasing on (-1,1). Since f''(x) = 6x, f''(x) > 0 for x > 0 and f''(x) < 0 for x < 0. Thus, f is concave upward for x > 0 and concave downward for x < 0. Also, a point of inflection occurs at x = 0. The graph of f appears in the figure below.

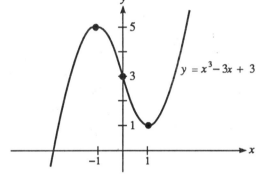

5. When $f(x) = x^3 - x^2$, we have

    $$f'(x) = 3x^2 - 2x = x(3x - 2)$$

    and

    $$f''(x) = 6x - 2$$

    The critical numbers of f are x = 0 and x = 2/3. Using the second derivative test
    $$f''(0) = -2 < 0 \text{ and } f''(2/3) = 6(2/3) - 2 = 2 > 0$$

    so f has a relative maximum at x = 0 and a relative minimum at x = 2/3. We analyze the sign of f' in the figure on the next page.

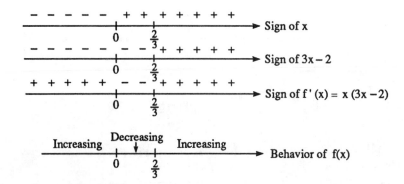

Thus, f is increasing when $x < 0$ and when $x > 2/3$, and f is decreasing when $0 < x < 2/3$. Since $f''(x) = 6x - 2$, $f''(x) > 0$ when $x > 1/3$ and $f''(x) < 0$ for $x < 1/3$. Thus f is concave upward for $x > 1/3$ and concave downward for $x < 1/3$. Also a point of inflection occurs at $x = 1/3$. The graph of f appears in the figure below.

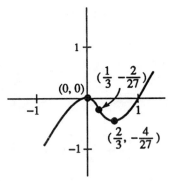

11. When $f(x) = x^4 + 4x^3$, we have

$$f'(x) = 4x^3 + 12x^2 = 4x^2(x+3)$$

and

$$f''(x) = 12x^2 + 24x = 12x(x + 2)$$

The critical numbers of f are $x = 0$ and $x = -3$. Using the second derivative test

$$f''(-3) = 12(-3)^2 + 24(-3) = 36 > 0$$

so f has a relative minimum at $x = -3$. Since $f''(0) = 0$, we cannot use the second derivative test. Since $f'(-1) > 0$ and $f'(1) > 0$, we have neither a relative maximum nor relative minimum at $x = 0$. We analyze the sign of f' in the figure on the next page.

4: *Applications of the Derivative* 89

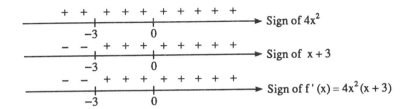

Thus f is increasing when x > -3 and decreasing when x < -3. Since f"(x) = 12x(x + 2), f"(x) > 0 when x > 0 or when x < -2 and f"(x) < 0 when -2 < x < 0. Thus f is concave upward for x > 0 or x < -2, and concave downward for -2 < x < 0. Also f has two points of inflection that occur at x = -2 and at x = 0. The graph of f appears in the figure below.

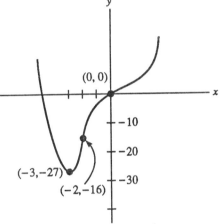

13. When $f(x) = 3x^{4/3}$, we have

$$f'(x) = 4x^{1/3}$$

and

$$f''(x) = \frac{4}{3}x^{-2/3} = \frac{4}{3(\sqrt[3]{x^2})}$$

The critical number of f is x = 0. Since f"(0) is undefined, we must use the first derivative test. When x < 0, f'(x) < 0 and when x > 0, f'(x) > 0. Thus f has a relative minimum at x = 0. We note that f is decreasing for x < 0 and increasing when x > 0. Since f"(x) > 0 for all non-zero values of x, f is always concave upward and there are no points of inflection. The graph of f appears in the figure on the next page.

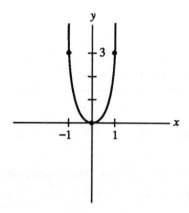

Exercise Set 4.6, (Page 240)

5. Given $f(x) = \dfrac{1}{x^2 + 1}$

   We note that the domain of this function is $(-\infty, \infty)$. Since there are no values of x for which the denominator equals zero, there are no vertical asymptotes. For all values of x, the numerator is less than or equal to the denominator. Hence $0 < f(x) \leq 1$ and the range of the function is the interval $(0, 1]$. Since $f(-x) = f(x)$, the graph is symmetric with respect to the y axis.

   $$\text{Now } \lim_{x \to \pm\infty} \frac{1}{x^2+1} = \lim_{x \to \pm\infty} \frac{\frac{1}{x^2}}{1 + \frac{1}{x^2}} = \frac{0}{1+0} = 0$$

   The line $y = 0$ is a horizontal asymptote of the function. To find relative extrema, we find

   $$f'(x) = -1(x^2+1)^{-2}(2x) = \frac{-2x}{(x^2+1)^2}$$

   Setting $f'(x) = 0$, we have $\dfrac{-2x}{(x^2+1)^2} = 0$ or $-2x = 0$ and thus $x = 0$ is a critical number. If $x < 0$, $f'(x) > 0$ and if $x > 0$, $f'(x) < 0$. Thus at $x = 0$, f has a relative maximum.

   To discuss concavity, we find

   $$f''(x) = (-2x)(-2)(x^2+1)^{-3}(2x) - 2(x^2+1)^{-2}$$

4: Applications of the Derivative 91

or

$$\frac{8x^2 - 2(x^2+1)}{(x^2+1)^3} = \frac{6x^2-2}{(x^2+1)^3}$$

Setting $f''(x) = 0$, we have $6x^2 = 2$, $x^2 = 1/3$, $x = \pm\sqrt{1/3}$

When $x < -\sqrt{1/3}$ or $x > \sqrt{1/3}$, $f''(x) > 0$, and thus f is concave upward. When $-\sqrt{1/3} < x < \sqrt{1/3}$, $f''(x) < 0$, and thus f is concave downward. Thus, at $x = \pm\sqrt{1/3}$, f has points of inflection. We make a table of values and sketch the graph below.

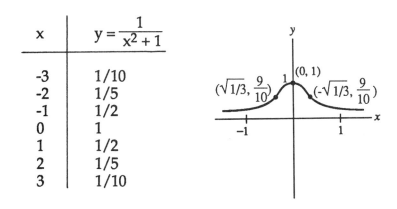

x	$y = \dfrac{1}{x^2+1}$
-3	1/10
-2	1/5
-1	1/2
0	1
1	1/2
2	1/5
3	1/10

7. Given $f(x) = \dfrac{1}{x-3}$

Since $f(x)$ is not defined at the value $x = 3$, the line $x = 3$ is a vertical asymptote.

$$\lim_{x \to \pm\infty} f(x) = \lim_{x \to \pm\infty} \frac{1}{x-3} = \lim_{x \to \pm\infty} \frac{\frac{1}{x}}{1-\frac{3}{x}} = \frac{0}{1-0} = 0$$

Thus the line $y = 0$ is a horizontal asymptote.

The graph is not symmetric with respect to the y-axis since $f(-x) \neq f(x)$.

Now $f'(x) = -1(x-3)^{-2} = \dfrac{-1}{(x-3)^2}$

There are no critical numbers of f, and thus no relative extrema. Since f'(x) < 0, the function is decreasing on (-∞, 3) and (3, ∞).

$$f''(x) = \frac{2}{(x-3)^3}$$

For x > 3, f''(x) > 0, and the graph is concave upward. For x < 3, f''(x) < 0, and the graph is concave downward. We make a table of values and sketch the graph below.

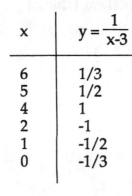
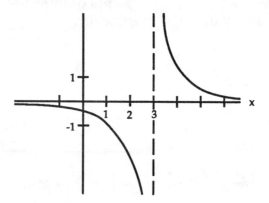

x	$y = \frac{1}{x-3}$
6	1/3
5	1/2
4	1
2	-1
1	-1/2
0	-1/3

13. Given $f(x) = \frac{1}{\sqrt{x}}$

The domain of this function is x > 0. The line x = 0 is a vertical asymptote of the function since $\lim_{x \to 0^+} f(x) = +\infty$. Note that $\lim_{x \to 0^-} f(x)$ does not exist. Since f(x) > 0, the graph lies above the x-axis. Also,

$$\lim_{x \to \pm\infty} f(x) = \lim_{x \to \pm\infty} \frac{1}{\sqrt{x}} = 0.$$

Thus the line y = 0 is a horizontal asymptote. The graph is not symmetric with respect to the y-axis, since values of x < 0 are not in the domain of f.

$$\text{Now } f'(x) = \frac{-1}{2}x^{-3/2} = \frac{-1}{2\sqrt{x^3}}$$

There are no critical numbers of f, and thus no relative extrema. Since f'(x) < 0 for all x in the domain of f, the function f(x) is decreasing on (0, ∞).

4: Applications of the Derivative 93

For x > 0, $f''(x) = \frac{3}{4}x^{-5/2} = \frac{3}{4\sqrt{x^5}}$

Since $f''(x) > 0$ for $x > 0$, the graph is concave upward.

We make a table of values and sketch the graph below

x	$y = \frac{1}{\sqrt{x}}$
1	1
4	1/2
9	1/3
16	1/4

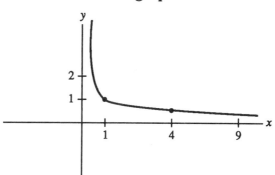

Exercise Set 4.7, (Page 254)

5. Let x denote the length (in meters) of fencing along one side of the grazing field perpendicular to the river. Then, the amount of fence parallel to the river is 600 - 2x (See the figure below).

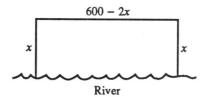

The area of the grazing field is

$A(x) = x(600 - 2x) = 600x - 2x^2$

Since both sides of the rectangle have positive length, we must have $x > 0$ and $600 - 2x > 0$. Thus, the problem reduces to finding the maximum value of A(x) subject to $0 < x < 300$.

Note that $A'(x) = 600 - 4x = 4(150 - x)$. Thus, $A'(x) > 0$ for $0 < x < 150$ and $A'(x) < 0$ for $150 < x < 300$. The area is then increasing for $0 < x < 150$ and decreasing for $150 < x < 300$. The maximum area occurs when x = 150. The dimensions of the corresponding rectangle are 150 meters and 600 - 2(150) = 300 meters.

9. Let y denote the average yield per tree (in bushels) and x denote the number of trees planted on the orchard. Now

$$y = \begin{cases} 60 & \text{if } 0 < x \leq 80 \\ 60 - 2(x - 80) & \text{if } x > 80 \end{cases}$$

Since $60 - 2(x - 80) = 220 - 2x = 2(110 - x)$ we must also have $x < 110$. The total yield of the orchard is the number of trees times the average yield per tree. Hence, the total yield is

$$T(x) = xy = \begin{cases} 60x & \text{if } 0 < x \leq 80 \\ 220x - 2x^2 & \text{if } 80 < x < 110 \end{cases}$$

Now

$$T'(x) = \begin{cases} 60 & \text{if } 0 < x < 80 \\ 220 - 4x & \text{if } 80 < x < 110 \end{cases}$$

Note that $T'(x) > 0$ for $0 < x < 80$ and $T'(x) < 0$ for $80 < x < 110$. Thus, the total yield is increasing for $0 < x < 80$ and decreasing for $80 < x < 110$. The maximum total yield occurs when 80 trees are planted.

When $x = 80$, the average yield per tree will be 60 bushels, and the total yield of the orchard will be $60(80) = 4800$ bushels.

15. Let x be the lot size. The average number of boats on hand during the year is $x/2$. The annual holding cost $H(x)$ is given by

$H(x) = $ (annual holding cost per boat) $\cdot$ (average number of boats)

$H(x) = 200(x/2) = 100x$

The annual reorder cost $O(x)$ is given by

$$O(x) = (\text{cost per order}) \cdot (\text{number of orders})$$

$$= (\text{cost per order}) \cdot \left(\frac{\text{number of boats sold per year}}{\text{lot size}}\right)$$

$$= (30)(1920/x)$$

The annual inventory cost $K(x)$ is

$$K(x) = H(x) + O(x)$$

$$K(x) = 100x + \frac{57600}{x}$$

We want to minimize $K(x)$ over the interval $[1, 1920]$.

$$K'(x) = 100 - \frac{57600}{x^2}$$

The critical numbers for $K(x)$ are obtained from solving

$$100x^2 = 57600$$
$$x^2 = 576$$
$$x^2 = \pm 24$$

Only the value $x = 24$ is in the interval $[1, 1920]$. We evaluate $K(x)$ at the critical number $x = 24$ and at the endpoints.

$$K(1) = 100 + 57600 = 57700$$
$$K(24) = 2400 + 2400 = 4800$$
$$K(1920) = 19200 + 30 = 19230$$

The minimum total annual inventory cost is $4800 when the economic ordering quantity is 24 boats. Thus the dealer should place

$$\frac{1920}{24} = 80 \text{ orders}$$

of 24 boats each. We note that the $10 per boat charge does not play a role in the solution of this problem.

21. 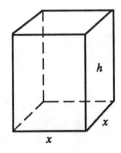 The volume of the box is the product of the length times the width times the height. Since the box has a square base each of whose sides is x and a height h, the volume $V = x^2 h = 96$, or $h = \dfrac{96}{x^2}$ (1)

The cost $C = 3x^2$ $+ 1(xh)$ 4
    (cost of base)   (cost of each side)   (number of sides)

$= 3x^2 + 4xh$.

Substituting for h from (1), we have

$$C(x) = 3x^2 + 4x\left(\dfrac{96}{x^2}\right) = 3x^2 + \dfrac{384}{x}$$

We wish to find the minimum of C(x). Now

$$C'(x) = 6x - \dfrac{384}{x^2}$$

Setting $C' = 0$, and solving for x yields

$$6x = \dfrac{384}{x^2} \text{ or } x^3 = \dfrac{384}{6} = 64$$

Thus $x = 4$.

We note that

$$C''(x) = 6 + \dfrac{768}{x^3}$$

and $C''(4) > 0$. Thus we have a minimum at $x = 4$. The minimum cost will be $C(4) = 3(4)^2 + \dfrac{384}{4} = 48 + 96 = \$144$.

25. Since the printed matter occupies 128 square inches, we have $xy = 128$ so that $y = 128/x$. The dimensions of the page are $x + 2$ and $y + 4$. Hence, the area of the page is

$$A(x) = (x + 2)(y + 4) = (x + 2)\left[\dfrac{128}{x} + 4\right] = 136 + 4x + \dfrac{256}{x}$$

We wish to minimize this area. We have

$$A'(x) = 4 - \frac{256}{x^2} = 4\frac{x^2 - 64}{x^2} \text{ and } A''(x) = 512/x^3$$

Note that $A'(8) = 0$ and $A''(8) = 1 > 0$ so that $A(x)$ has a relative minimum at $x = 8$. The corresponding value of y is $y = 128/8 = 16$. Thus, the printed matter occupies a rectangle with dimensions 8 by 16 inches. The page is 10 inches by 20 inches.

27. Note that the amount of underwater pipeline (in miles) is $\sqrt{4 + x^2}$ and the amount of pipeline along the shoreline is $8 - x$. The total cost (in dollars) of the pipeline is

$$C(x) = 1000\sqrt{4 + x^2} + 800(8 - x)$$

Since $0 \leq x \leq 8$, the problem reduces to minimizing $C(x)$ over this interval.

Now

$$C'(x) = 500(4 + x^2)^{-1/2}(2x) + 800(-1) = \frac{1000x}{\sqrt{4 + x^2}} - 800$$

and

$$C''(x) = \frac{\sqrt{4 + x^2}(1000) - 1000x\frac{1}{2}\frac{1}{\sqrt{4+x^2}}2x}{4 + x^2}$$

$$= \frac{1000}{(4 + x^2)\sqrt{4 + x^2}}(4 + x^2 - x^2) = \frac{4000}{[4 + x^2]^{3/2}} > 0$$

We set $C'(x) = 0$ obtaining

$$1000x = 800\sqrt{4 + x^2}$$
$$5x/4 = \sqrt{4 + x^2}$$
$$25x^2/16 = 4 + x^2$$
$$9x^2/16 = 4$$
$$x = \pm 8/3$$

98 Study Guide

By substitution we find that x = -8/3 should be rejected and that C'(8/3) = 0. Since C"(8/3) > 0, the cost is minimized at x = 8/3. Thus, the point D should be located 1/3 of the way from C to A.

<u>Exercise Set 4.8, (Page 264)</u>

1. Since we are approximating a square root, we let $f(x) = \sqrt{x}$. We also choose $x_0 = 36$ and $h = 1$. We use

    $$f(x_0 + h) = f(x_0) + f'(x_0)h$$

    In our problem,

    $$f(37) = \sqrt{37} = f(36) + f'(36) \cdot 1$$

    Now, $f'(x) = 1/(2\sqrt{x})$ so that $f'(36) = 1/(2\sqrt{36}) = 1/12$. Thus

    $$\sqrt{37} = \sqrt{36} + \frac{1}{12} = 6\frac{1}{12} = \frac{73}{12} = 6.083$$

11. When $f(x) = [1 + 2x^2]^{1/2}$, we have

    $$f'(x) = \frac{1}{2}[1 + 2x^2]^{-1/2} \cdot (4x) = 2x[1 + 2x^2]^{-1/2}$$

    Now $df = f'(x_0)dx$. When $x_0 = 2$ and $dx = 0.2$ we have

    $$df = f'(2)(0.2)$$
    $$= 2(2)[1 + 2(2)^2]^{-1/2}(0.2)$$
    $$= \frac{0.8}{\sqrt{9}} = \frac{8}{30} = \frac{4}{15} = 0.267$$

15. We have

    $$\frac{dy}{dx} = 6x^2 - 6x + 5$$

    Hence $dy = (6x^2 - 6x + 5)dx$

17. We have

4: *Applications of the Derivative* 99

$$\frac{dy}{dx} = \frac{(x+1)(1) - (x-1)(1)}{(x+1)^2} = \frac{2}{(x+1)^2}$$

Hence

$$dy = \frac{2}{(x+1)^2} dx$$

19. We have

$$\frac{dy}{dx} = 5(x^2+1)^4(2x) = 10x(x^2+1)^4$$

Hence,

$$dy = 10x(x^2+1)^4 \, dx$$

23. We have

$$\frac{dV}{dr} = \frac{4}{3}\pi(3r^2) = 4\pi r^2$$

so

$$dV = 4\pi r^2 \, dr$$

We are given $r_0 = 2$ and $dr = -0.2$. Then

$$dV = 4\pi(2)^2(-0.2) = -3.2\pi = -10.0531$$

Thus, when radiation reduces the radius from 2 cm to 1.8 cm, the volume of the tumor decreases by approximately 10.05 cubic centimeters.

29. We have

$$\frac{dT}{dx} = \frac{1}{2}x^{-1/2}$$

so

$$dT = \frac{1}{2\sqrt{x}} dx$$

We are given $x = 1600$ and $dx = 50$. Thus

$$dT = \frac{1}{2\sqrt{1600}}(50) = \frac{50}{2(40)} = \frac{5}{8} = .625$$

When a person's monthly salary increases from $1600 to $1650, the additional monthly tax due is 63 cents.

Exercise Set 4.9, (Page 272)

5. To find $\sqrt[4]{100}$, we must solve the equation $x^4 - 100 = 0$. Let $f(x) = x^4 - 100$, and try to solve the equation $f(x) = 0$. We note that $f'(x) = 4x^3$.

   The Newton-Raphson algorithm uses the formula

   $$x_{n+1} = x_n - \frac{f(x_n)}{f'(x_n)}$$

   or

   $$x_{n+1} = x_n - \frac{x_n^4 - 100}{4x_n^3}$$

   Since $f(3) = -19 < 0$ and $f(4) = 156 > 0$, the function must have a zero between 3 and 4. Since the value $x = 3$ seems to be closer to the zero, we shall choose it as our first guess.

   Successive guesses are obtained from the above formula.

   $$x_2 = 3 - \frac{3^4 - 100}{4(3)^3} = 3 - \frac{-19}{108} = 3.176$$

   $$x_3 = 3.176 - \frac{[(3.176)^4 - 100]}{4(3.176)^3} = 3.176 - \frac{1.747}{128.145} = 3.162$$

   $$x_4 = 3.162 - \frac{[(3.162)^4 - 100]}{4(3.162)^3} = 3.162 - \frac{-.035}{126.458} = 3.162$$

   Since x_3 and x_4 agree to the nearest thousandth, we are satisfied that we have found $\sqrt[4]{100}$ to the desired accuracy. Thus $\sqrt[4]{100} \approx 3.16$.

7. Let $f(x) = x^3 - x - 5$. We seek a positive number that solves the equation $f(x) = 0$. We note that $f'(x) = 3x^2 - 1$. The Newton-Raphson algorithm formula is

4: Applications of the Derivative 101

$$x_{n+1} = x_n - \frac{f(x_n)}{f'(x_n)}$$

or

$$x_{n+1} = x_n - \frac{x_n^3 - x_n - 5}{3x_n^2 - 1}$$

Since $f(1) = -5 < 0$ and $f(2) = 1 > 0$, the function must have a zero between 1 and 2 from the intermediate value property. Since the value $x = 2$ seems to be closer to the zero, we shall choose it as our first guess. Successive guesses are obtained from the above formula.

$$x_2 = 2 - \frac{(2)^3 - 2 - 5}{3(2)^2 - 1} = 2 - \frac{1}{11} = 1.909$$

$$x_3 = 1.909 - \frac{[(1.909)^3 - (1.909) - 5]}{3(1.909)^2 - 1} = 1.909 - \frac{.048}{9.933} = 1.904$$

$$x_4 = 1.904 - \frac{[(1.904)^3 - (1.904) - 5]}{3(1.904)^2 - 1} = 1.904 - \frac{-.002}{9.876} = 1.904$$

Since x_3 and x_4 agree to the nearest thousandth, we have $x = 1.90$ to the nearest hundredth. Thus 1.90 is an approximate root of the equation $x^3 - x - 5 = 0$.

9. Squaring both sides gives

$$(x + 1)^2 = x + 2$$
$$x^2 + 2x + 1 - x - 2 = 0$$
$$x^2 + x - 1 = 0$$

Let $f(x) = x^2 + x - 1$. We seek a positive root of the equation $f(x) = 0$. We note that $f'(x) = 2x + 1$. The Newton-Raphson algorithm formula is

$$x_{n+1} = x_n - \frac{f(x_n)}{f'(x_n)}$$

or

$$x_{n+1} = x_n - \frac{x_n^2 + x_n - 1}{2x_n + 1}$$

Since $f(0) = -1 < 0$ and $f(1) = 1 > 0$, the function must have a zero between 0 and 1. Both values seem to be "equidistant" from the root, so we choose $x = .5$ as our first guess. Successive guesses are obtained from the above formula.

$$x_1 = .5 - \frac{(.5^2 + .5 - 1)}{2(.5) + 1} = .5 - \frac{-.25}{2} = .625$$

$$x_2 = .625 - \frac{(.625^2 + .625 - 1)}{2(.625) + 1} = .625 - \frac{.016}{2.25} = .618$$

$$x_3 = .618 - \frac{(.618^2 + .618 - 1)}{2(.618) + 1} = .618 - \frac{(-.0000076)}{2.236} = .618$$

Since x_2 and x_3 agree to the nearest thousandth, we have $x = .62$ to the nearest hundredth. Thus $x = .62$ is an approximate root of the equation $x + 1 = \sqrt{x + 2}$.

15. Given $s = 10t^2 - t^3$ for $0 \le t \le 10$. We want to find t for $s = 80$. Thus we wish to solve the equation

$$80 = 10t^2 - t^3 \text{ or } t^3 - 10t^2 + 80 = 0.$$

Let $f(t) = t^3 - 10t^2 + 80$. We seek a positive value of t between 0 and 10 such that $f(t) = 0$. We note that $f'(t) = 3t^2 - 20t$. The Newton-Raphson algorithm formula is

$$t_{n+1} = t_n - \frac{f(t_n)}{f'(t_n)}$$

or

$$t_{n+1} = t_n - \frac{t_n^3 - 10t_n^2 + 80}{3t_n^2 - 20t_n}$$

Since $f(3) = 17 > 0$ and $f(4) = -16 < 0$, the function must have a zero between 3 and 4. Both values seem to be "equidistant" from the root, so we choose $t = 3.50$ as a first guess. Successive guesses are obtained from the above formula.

$$t_2 = 3.50 - \frac{[3.50^3 - 10(3.50)^2 + 80]}{3(3.50)^2 - 20(3.50)} = 3.5 - \frac{.375}{(-33.25)} = 3.51$$

4: *Applications of the Derivative* 103

$$t_3 = 3.51 - \frac{[3.51^3 - 10(3.51)^2 + 80]}{3(3.51)^2 - 20(3.51)} = 3.51 - \frac{.043}{(-33.239)} = 3.51$$

Since t_2 and t_3 agree to the nearest hundredth, we are satisfied that we have found $t = 3.5$ to the nearest tenth. Thus it will take approximately 3.5 seconds for the object to move 80 feet.

Review Exercises, (Page 274)

11. We have $f'(x) = 6x^2 - 4x^3$. Setting $f'(x) = 0$ gives $6x^2 - 4x^3 = 0$ or $2x^2(3 - 2x) = 0$. Thus $x = 0$ and $x = 3/2$ are the critical numbers of f. At these values f will have a horizontal tangent.

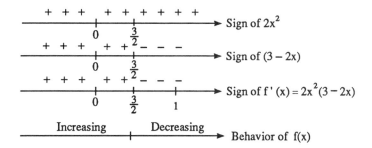

   f is increasing on $(-\infty, 3/2)$ and decreasing on $(3/2, \infty)$.

13. We have $f'(x) = 2 - 2x = 2(1 - x)$. Since $f'(x)$ exists for all values of x, we conclude that the critical numbers of f occur when $f'(x) = 0$. Thus, $x = 1$ is the only critical number of f. Note that $f'(x) > 0$ for $x < 1$ and $f'(x) < 0$ for $x > 1$. Since the sign of f' changes from + to - at $x = 1$, f has a relative maximum at $x = 1$.

19. Since $f(x) = 3x^{1/3} - 4x$ is continuous on the closed interval $[0,8]$, the Extreme Value Theorem guarantees an absolute maximum and an absolute minimum value for f on $[0,8]$. These must occur either at critical numbers or at endpoints.

    We have

    $$f'(x) = x^{-2/3} - 4$$

    Setting $f'(x) = 0$ we obtain

    $$x^{-2/3} = 4$$

    $$x^{2/3} = 1/4$$

    $$[x^{2/3}]^{3/2} = (1/4)^{3/2} = (\sqrt{1/4})^3 = 1/8$$

or

$$x = 1/8$$

Also, f'(x) does not exist at x = 0. Thus, x = 0 and x = 1/8 are the only critical numbers of f. We evaluate f at the critical numbers and at the endpoints obtaining

$$f(0) = 3(0)^{1/3} - 4(0) = 0$$
$$f(1/8) = 3(1/8)^{1/3} - 4(1/8) = 1$$
$$f(8) = 3(8)^{1/3} - 4(8) = -26$$

The absolute maximum of f on [0,8] is 1 which occurs at x = 1/8. The absolute minimum of f on [0, 8] is -26 which occurs at x = 8.

23. When $f(x) = 2x^3 + 3x^2 - 12x + 3$, we have

$$f'(x) = 6x^2 + 6x - 12$$

and

$$f''(x) = 12x + 6 = 6(2x + 1)$$

Note that f''(x) < 0 for x < -1/2 and f''(x) > 0 for x > -1/2. Hence, f is concave downward for x < -1/2 and concave upward for x > -1/2. Since f is defined at x = -1/2 and f'' changes sign at x = -1/2, f has an inflection point at x = -1/2. The corresponding y value at the inflection point is

$$f(-\frac{1}{2}) = 2(-\frac{1}{2})^3 + 3(-\frac{1}{2})^2 - 12(-\frac{1}{2}) + 3 = \frac{19}{2}$$

25. When $f(x) = x^3 - 12x + 8$, we have

$$f'(x) = 3x^2 - 12 = 3(x^2 - 4) = 3(x + 2)(x - 2)$$

and

$$f''(x) = 6x$$

The critical numbers of f are x = -2 and x = 2. Now

$$f''(-2) = 6(-2) = -12 < 0 \text{ and } f''(2) = 6(2) = 12 > 0$$

so that f has a relative maximum at x = -2 and a relative minimum at x = 2. The relative maximum is f(-2) = 24 and the relative minimum is f(2) = -8.

37. (a) Since we are approximating a square root, we let $f(x) = x^{1/2}$. We also choose $x_0 = 49$ and $h = -.7$. Now $f'(x) = \frac{1}{2}x^{-1/2}$ so that

$$f'(x_0) = \frac{1}{2}(49)^{-1/2} = \frac{1}{2\sqrt{49}} = \frac{1}{2(7)} = \frac{1}{14}$$

and we note that $f(x_0) = 49^{1/2} = 7$. Thus we have

$$f(x_0 + h) = (48.3)^{1/2} = f(x_0) + f'(x_0)h$$

$$= 7 + \frac{1}{14}(-.7) = 7 - \frac{1}{20} = \frac{139}{20} = 6.95$$

(b) Since we are approximating a two-thirds power, we let $f(x) = x^{2/3}$. We also choose $x_0 = 27$ and $h = 0.04$. Now, $f'(x) = \frac{2}{3}x^{-1/3}$, so that

$$f'(x_0) = \frac{2}{3}(27)^{-1/3} = \frac{2}{3} \cdot \frac{1}{\sqrt[3]{27}} = \frac{2}{3} \cdot \frac{1}{3} = \frac{2}{9}$$

and noting that

$$f(x_0) = (27)^{2/3} = (\sqrt[3]{27})^2 = 3^2 = 9$$

we have

$$f(x_0 + h) = (27.04)^{2/3} = f(x_0) + f'(x_0)h$$

$$= 9 + \frac{2}{9}(0.04) = 9 + \frac{8}{900} = \frac{2027}{225} = 9.009$$

39. We have

$$\frac{dy}{dx} = 3x^2 - 1$$

so

$$dy = (3x^2 - 1)\,dx$$

43. The marginal profit is given by

106 Study Guide

$$P'(x) = \frac{1}{3}x^3 - \frac{7}{2}x^2 + 12x$$

The marginal profit is increasing only when its derivative P''(x) is positive. Note

$$P''(x) = x^2 - 7x + 12 = (x-3)(x-4)$$

We analyze the sign of P'' in the next figure. Since x represents the level of production, the sign analysis involves only $x \geq 0$.

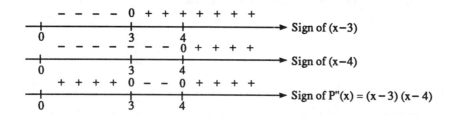

Thus, the marginal profit is increasing on (0,3) and (4,∞).

47. Let r denote the radius of the barrel (in feet) and h denote its height (in feet). The volume V is fixed at 32 cubic feet; thus

$$32 = \pi r^2 h$$

The surface area A is given by

$$A = 2\pi r^2 + 2\pi r h$$

To minimize the amount of material required we minimize surface area. First we write A as a function of the one variable r by noting that $h = 32/(\pi r^2)$ and substituting into the surface area formula. Thus, we minimize

$$A(r) = 2\pi r^2 + 2\pi r \frac{32}{\pi r^2} = 2\pi r^2 + \frac{64}{r}$$

for r > 0. Now

$$A'(r) = 4\pi r - \frac{64}{r^2} = \frac{4}{r^2}(\pi r^3 - 16)$$

and
$$A''(r) = 4\pi + \frac{128}{r^3}$$

The only critical number of A is the solution of
$$\pi r^3 - 16 = 0$$
or
$$\pi r^3 = 16$$
$$r = \sqrt[3]{\frac{16}{\pi}} = 1.72 \text{ ft.}$$

Since $A''(r) > 0$ for all values of $r > 0$, A has a relative minimum at $r = \sqrt[3]{\frac{16}{\pi}}$. The corresponding height is

$$h = \frac{32}{\pi r^2} = \frac{32}{\pi(\sqrt[3]{16/\pi})^2} = 2[\frac{16}{\pi}]/[\frac{16}{\pi}]^{2/3} = 2[\frac{16}{\pi}]^{1/3} = 3.44 \text{ ft.}$$

49. Since $V = \frac{4}{3}\pi r^3$, we have

$$\frac{dV}{dr} = \frac{4}{3}\pi(3r^2)$$

or
$$dV = 4\pi r^2 \, dr$$

We have $r_0 = 10$ and $dr = .005$. Thus
$$dV = 4\pi(10)^2 (.005) = 2\pi = 6.2832$$

When the radius increases from 10 to 10.005 feet, the volume will increase by approximately 6.28 cubic feet.

Chapter Test, (Page 276)

1. We find the derivative of f and determine the open intervals where f'(x) is positive and where it is negative. We have

$$f'(x) = 3x^2 + 6x - 9 = 3(x^2 + 2x - 3) = 3(x - 1)(x + 3)$$

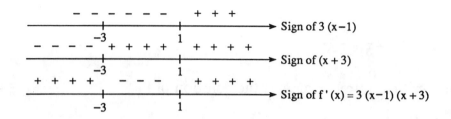

(a) Thus f is increasing on (-∞, -3) and on (1, ∞).

(b) f is decreasing on (-3, 1).

(c) The graph of f has horizontal tangents at x = -3 and at x = 1 since f'(-3) = f'(1) = 0. Since f(-3) = 20, and f(1) = -12, the points are (-3, 20) and (1, -12).

(d) Since the sign of f' changes from + to - at x = -3, we have a relative maximum at (-3, 20). Since the sign of f' changes from - to + at x = 1, we have a relative minimum at (1, -12).

(e) We have f"(x) = 6x + 6 = 6(x + 1). We analyze the sign of f" in the figure below

```
    - - - -    + + + + +
  ─────────────┼─────────────→ Sign of f" (x) = 6 (x+1)
              -1
```

Thus the graph of f is concave upward for x > -1.

(f) Since the second derivative changes sign at x = -1, the graph has a point of inflection. Now f(-1) = 4, so the point of inflection is (-1,4).

3. When $g(x) = x + \frac{1}{x}$, we have

$$g'(x) = 1 - \frac{1}{x^2}$$

and

$$g''(x) = \frac{2}{x^3}$$

4: Applications of the Derivative 109

Setting $g'(x) = 0$, we have

$$1 = \frac{1}{x^2} \text{ or } x^2 = 1$$

and the critical numbers of $g(x)$ are $x = 1$ and $x = -1$. From the second derivative test, $g''(1) > 0$, and $g''(-1) < 0$. Thus $g(x)$ has a relative minimum at $x = 1$ and a relative maximum at $x = -1$.

5. When $y = \sqrt{5x^2 + 6} = (5x^2 + 6)^{1/2}$

$$\frac{dy}{dx} = \frac{1}{2}(5x^2 + 6)^{-1/2}(10x)$$

$$= \frac{5x}{(5x^2 + 6)^{1/2}} = \frac{5x}{\sqrt{5x^2 + 6}}$$

So $dy = \dfrac{5x}{\sqrt{5x^2 + 6}} dx$

7. When $k(t) = \dfrac{0.1t}{4t^2 + 3t + 9}$, we have

$$k'(t) = \frac{(4t^2 + 3t + 9)(0.1) - (0.1t)(8t + 3)}{(4t^2 + 3t + 9)^2}$$

Since the equation $4t^2 + 3t + 9 = 0$ has no real roots, there are no values of t for which the denominator of $k'(t)$ is zero. Thus the only critical numbers of $k(t)$ will be those values of t for which the numerator of $k'(t)$ is zero.

Thus

$$.4t^2 + .3t + .9 - .8t^2 - .3t = 0$$
$$-.4t^2 = -.9$$
$$t^2 = 9/4$$
$$t = \pm 3/2$$

At $t = 1$, $k'(1) > 0$ and at $t = 2$, $k'(2) < 0$. Thus $k(t)$ has a maximum at $t = 3/2 = 1.5$ hours.

9. Given $f(x) = x^3 - x - 20$

(a) Now $f(2) = (2)^3 - 2 - 20 = -14 < 0$
and $f(3) = (3)^3 - 3 - 20 = 4 > 0$

Thus from the intermediate value property for continuous functions, there must exist a number c between 2 and 3 such that $f(c) = 0$.

(b) The Newton Raphson algorithm formula is

$$x_{n+1} = x_n - \frac{f(x_n)}{f'(x_n)}$$

Since $f'(x) = 3x^2 - 1$, we have

$$x_{n+1} = x_n - \frac{x_n^3 - x_n - 20}{3x_n^2 - 1}$$

Letting $x_1 = 3$, we have

$$x_2 = 3 - \frac{(3)^3 - (3) - 20}{3(3)^2 - 1} = 3 - \frac{4}{26} = 2.8462$$

$$x_3 = 2.8462 - \frac{(2.8462)^3 - (2.8462) - 20}{3(2.8462)^2 - 1} = 2.8462 - \frac{.2105}{23.3026} = 2.8372$$

5 The Exponential and Logarithmic Functions

Key Ideas for Review

* An exponential function is one defined by

    $f(x) = a^x \quad a>0, a \neq 1.$

* $e = \lim\limits_{n \to \infty} (1 + \frac{1}{n})^n \approx 2.718$

* **The exponential function** is $f(x) = e^x$

* If $f(x) = a^x$, with $a > 1$, then

    - the domain of f is $(-\infty, +\infty)$.
    - $f(x) > 0$ for all x.
    - f is an increasing function for all x
    - the graph of f passes through (0,1).
    - $\lim\limits_{x \to \infty} f(x) = \infty$.
    - the graph of f is concave upward for all x.

* If $f(x) = a^{-x}$, with $a > 1$, then

    - the domain of f is $(-\infty, +\infty)$.
    - $f(x) > 0$ for all x.
    - f is a decreasing function for all x.
    - the graph of f passes through (0, 1).
    - $\lim\limits_{x \to \infty} f(x) = 0$
    - the graph of f is concave upward for all x.

* If a principal P is invested at an annual rate r, compounded k times a year, and S is the value of the investment after n conversion periods then

    $$S = P(1 + \frac{r}{k})^n \text{ and } P = S(1 + \frac{r}{k})^{-n}$$

* If a principal P is invested at an annual rate r, compounded continuously, and S is the value of the investment after t years, then

    $S = Pe^{rt}$ and $P = Se^{-rt}$

* $y = \log_a x$ if and only if $x = a^y$. The base a is always a positive number different from 1.

* The **natural logarithm** of x is $\ln x = \log_e x$.

* The **common logarithm** of x is $\log x = \log_{10} x$.

* $a^{\log_a x} = x$; $\log_a(a^x) = x$. $e^{\ln x} = x$; $\ln e^x = x$.

* $\log_a a = 1$; $\log_a 1 = 0$. $\ln e = 1$; $\ln 1 = 0$.

* If $f(x) = \log_a x$, with $a > 1$, then

    - the domain of f is $(0, +\infty)$ and the range of f is $(-\infty, +\infty)$.
    - $f(x) > 0$ if $x > 1$; $f(x) < 0$ if $x < 1$.
    - f is an increasing function for all x.
    - the graph of f passes through the point (1, 0).
    - $\lim_{x \to +\infty} f(x) = +\infty$; $\lim_{x \to 0^+} f(x) = -\infty$.
    - the graph of f is the reflection of the graph of $y = a^x$ across the line $y = x$.
    - the graph of f is concave downward.

* $\log_a xy = \log_a x + \log_a y$.

* $\log_a \frac{x}{y} = \log_a x - \log_a y$. $\log_a \frac{1}{x} = -\log_a x$.

* $\log_a x^k = k \log_a x$.

* $\log_a x = \frac{\ln x}{\ln a} = \frac{\log_b x}{\log_b a}$.

* $\frac{d}{dx}[e^x] = e^x$.

5: The Exponential and Logarithmic Functions 113

* $\frac{d}{dx}[e^{g(x)}] = e^{g(x)} g'(x)$.

* $\frac{d}{dx}[e^u] = e^u \frac{du}{dx}$.

* $\frac{d}{dx}[a^x] = a^x \ln a$; $\frac{d}{dx}[a^{g(x)}] = a^{g(x)} g'(x) \ln a$.

* $\frac{d}{dx}[\ln x] = \frac{1}{x}$ (assuming $x > 0$).

* $\frac{d}{dx}[\ln g(x)] = \frac{1}{g(x)} g'(x)$ (assuming $g(x) > 0$).

* $\frac{d}{dx}[\log_b x] = \frac{1}{x \ln b}$ (assuming $x > 0$).

* $\frac{d}{dx}[\log_b g(x)] = \frac{g'(x)}{g(x) \ln b}$ (assuming $g(x) > 0$).

* If $Q = Q_0 e^{kt}$, then $dQ/dt = kQ$.

* The solution to the differential equation $dQ/dt = kQ$ where k is a constant, is $Q = Q_0 e^{kt}$, where Q_0 is the constant $Q(0)$.

* Q is growing exponentially if $Q = Q_0 e^{kt}$, $k > 0$. The amount initially present is Q_0, and the constant k is called the growth constant (or growth rate).

* Q is decaying exponentially if $Q = Q_0 e^{-kt}$, $k > 0$. The amount initially present is Q_0, and the constant k is called the decay constant (or rate of decay).

* If $Q = Q_0 e^{kt}$, then the number $T = \frac{\ln 2}{|k|}$ is called the doubling time if $k > 0$, or the half-life of $k < 0$.

Exercise Set 5.1, (Page 290)

3. To sketch $f(x) = 4^x$, we form a table of values and plot the corresponding points.

114 Study Guide

x	-2	-1.5	-1	-0.5	0	0.5	1	1.5	2
$f(x) = 4^x$	1/16	1/8	1/4	1/2	1	2	4	8	16

The sketch appears in the figure below.

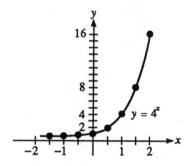

7. To sketch $f(x) = (1/5)^x$, we form a table of values and plot the corresponding points.

x	-2	-1	0	1	2
$f(x) = (1/5)^x$	25	5	1	1/5	1/25

The sketch appears in the figure below.

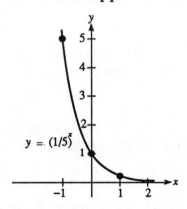

17. (a) We are given P = $6000, r = 0.06, k = 1, and n = 4. Then

$$i = \frac{r}{k} = \frac{0.06}{1} = 0.06$$

and the future value S of the investment is

$$S = P(1 + r)^n = 6000(1 + 0.06)^4 = \$7574.88$$

The interest is $7574.88 - $6000 = $1574.88.

(b) We are given P = $6000, r = 0.06, k = 12, and n = (12)(4) = 48. Then

$$i = \frac{r}{k} = \frac{0.06}{12} = 0.005$$

and the future value S of the investment is

$$S = P(1 + r)^n = 6000(1 + 0.005)^{48} = \$7622.94$$

The interest is $7622.94 - $6000 = $1622.94.

(c) We are given P = $6000, r = 0.06, k = 4, and n = (4)(4) = 16. Then

$$i = \frac{r}{k} = \frac{0.06}{4} = 0.015$$

and the future value S of the investment is

$$S = P(1 + r)^n = 6000(1 + 0.015)^{16} = \$7613.94$$

The interest is $7613.94 - $6000 = $1613.94.

(d) We are given P = $6000, r = 0.06, k = 2 and n = 2(4) = 8. Then

$$i = \frac{r}{k} = \frac{0.06}{2} = 0.03$$

and the future value S of the investment is

$$S = P(1 + r)^n = 6000(1 + 0.03)^8 = \$7600.62$$

The interest is $7600.62 - $6000 = $1600.62.

(e) We are given P = 6000, r = 0.06, t = 4. The future value S of the investment is

$$S = Pe^{rt} = 6000e^{(0.06)(4)} = \$7627.49$$

The interest is $7627.49 - $6000 = $1627.49.

19. (a) We have P = $12,000, r = 0.09, k = 2, and n = (8)(2) = 16 conversions. The future value of the investment is

$$S = P(1 + \frac{r}{k})^n = 12,000(1 + \frac{0.09}{2})^{16} \approx \$24,268.44.$$

(b) We have P = $12,000, r = 0.09, t = 8. The future value S of the investment is

$$S = Pe^{rt} = 12{,}000e^{(0.09)(8)} = \$24{,}653.20.$$

25. (a) We have r = 0.08 and k = 2. The effective rate of interest is

$$V = (1 + \tfrac{r}{k})^k - 1 = (1 + \tfrac{0.08}{2})^2 - 1 = 0.0816$$

The effective rate is 8.16%.

(b) We have r = 0.08 and k = 4. The effective rate of interest is

$$V = (1 + \tfrac{r}{k})^k - 1 = (1 + \tfrac{0.08}{4})^4 - 1 = 0.0824$$

The effective rate is 8.24%.

(c) We have r = 0.08 and k = 12. The effective rate of interest is

$$V = (1 + \tfrac{r}{k})^k - 1 = (1 + \tfrac{0.08}{12})^{12} - 1 = 0.083$$

The effective rate is 8.30%.

(d) We have r = 0.08. The effective rate of interest is

$$V = e^r - 1 = e^{0.08} - 1 = 0.0833$$

The effective rate is 8.33%.

29. The population P of the bacteria colony is given by $P(t) = 2000e^{0.04t}$ where t is the number of days that the colony has been growing.

(a) After 10 days, we have t = 10 and

$$P(10) = 2000e^{(0.04)(10)} = 2984 \text{ bacteria.}$$

(b) After 80 days, we have t = 80 and

$$P(80) = 2000e^{(0.04)(80)} = 49{,}065 \text{ bacteria.}$$

Exercise Set 5.2, (Page 299)

3. (a) $\log_2 16 = 4$, since $2^4 = 16$
   (b) $\log_2 1/2 = -1$, since $2^{-1} = 1/2$
   (c) $\log_2 128 = 7$, since $2^7 = 128$
   (d) $\log_2 \sqrt[3]{32} = 5/3$, since $2^{5/3} = (2^5)^{1/3} = \sqrt[3]{32}$

7. (a) $\log 15 = \log (5)(3) = \log 5 + \log 3$

   $= 0.6990 + 0.4771$
   $= 1.1761$

   (b) $\log 5/3 = \log 5 - \log 3 = 0.6990 - 0.4771 = 0.2219$

   (c) $\log 3000 = \log (3)(1000) = \log 3 + \log 1000$

   $= 0.4771 + 3 = 3.4771$

   (d) $\log 1/500 = \log 1 - \log 500 = -\log 500$

   $= -\log (5)(100) = -[\log 5 + \log 100]$
   $= -[0.6990 + 2] = -2.6990$

   (e) $\log 75 = \log (25)(3) = \log 25 + \log 3$

   $= \log (5)^2 + \log 3 = 2 \cdot \log 5 + \log 3$
   $= 2(0.6990) + 0.4771 = 1.8751$

   (f) $\log \sqrt{45} = \log (45)^{1/2} = \frac{1}{2} \log 45$

   $= \frac{1}{2} \log (9)(5) = \frac{1}{2} [\log 9 + \log 5]$
   $= \frac{1}{2} [2 \cdot \log 3 + \log 5]$
   $= \log 3 + \frac{1}{2} \log 5$
   $= 0.4771 + \frac{1}{2} (0.6990) = 0.8266$

9. (a) Since $\log_4 x = 3$ is equivalent to $4^3 = x$, we see that $x = 64$.

   (b) Since $\log_5 x = -1$ is equivalent to $5^{-1} = x$, we see that $x = \frac{1}{5}$.

   (c) Since $\log_x 9 = 2$ is equivalent to $x^2 = 9$ and $x > 0$, we see that $x = 3$.

   (d) Since $\log_x 4 = -2$ is equivalent to $x^{-2} = 4$ or $1/x^2 = 4$, we have

$x^2 = 1/4$. Thus $x = 1/2$ because $x > 0$.

19. (a) $e^t = 4$
    $t = \ln 4$
    $t = 1.3863$

    (b) $500e^{2t} = 3000$
    $e^{2t} = 3000/500 = 6$
    $2t = \ln 6$
    $t = \ln 6/2 = 0.8959$

    (c) $200e^{0.2t} = 100$
    $e^{0.2t} = 100/200 = .5$
    $0.2t = \ln .5$
    $t = \ln .5/0.2 = -3.4657$

    (d) $50e^{-2t} = 60$
    $e^{-2t} = 60/50 = 1.2$
    $-2t = \ln 1.2$
    $t = \ln 1.2/-2 = -0.0912$

    (e) $100(1 - e^{-0.2t}) = 40$
    $1 - e^{-0.2t} = 40/100 = .4$
    $-e^{-0.2t} = .4 - 1 = -.6$
    $-0.2t = \ln .6$
    $t = \ln .6/-0.2 = 2.5541$

    (f) $4 \ln x = 5$
    $\ln x = 5/4 = 1.25$
    $x = e^{1.25} = 3.4903$

21. Given $\log_a u = b$, $\log_a v = c$ and $\log_a w = d$

    (a) $\log_a \frac{u^3}{v^2} = \log_a u^3 - \log_a v^2$
    $= 3 \log_a u - 2 \log_a v$
    $= 3b - 2c$

    (b) $\log_a \sqrt{\frac{u}{vw}} = \log_a (\frac{u}{vw})^{1/2} = \frac{1}{2} \log_a \frac{u}{vw}$
    $= \frac{1}{2}[\log_a u - \log_a vw]$
    $= \frac{1}{2}[\log_a u - (\log_a v + \log_a w)]$

5: *The Exponential and Logarithmic Functions* 119

$$= \frac{1}{2}[b - (c + d)] = \frac{1}{2}(b - c - d)$$

(c) $\log_a u \sqrt[3]{v} = \log_a uv^{1/3}$

$$= \log_a u + \log_a v^{1/3}$$
$$= b + \frac{1}{3}\log_a v$$
$$= b + \frac{1}{3}c$$

25. The future value of $8000 invested at 8% (r = 0.08) compounded continuously is

$$S = Pe^{rt} = 8000e^{0.08t}$$

When S = $20,000, we have

$$8000e^{0.08t} = 20,000$$

$$e^{0.08t} = 20,000/8000 = 2.5$$

$$0.08t = \ln 2.5$$

$$t = \frac{\ln 2.5}{0.08} = 11.4536$$

Thus, it takes approximately 11.45 years for $8000 to grow to $20,000 when it is invested at 8% compounded continuously.

31. We have N = 4000 + 500 ln(x + 2). When N = 6000 mopeds,

$$6000 = 4000 + 500 \ln(x + 2)$$
$$2000 = 500 \ln(x + 2)$$
$$4 = \ln(x + 2)$$
$$e^4 = x + 2$$
$$54.598 = x + 2$$
$$52.598 = x$$

Thus $52,598 of advertising money must be spent to sell 6000 mopeds.

Exercise Set 5.3, (Page 310)

7. $\frac{d}{dx}[e^{\sqrt{3x+2}}] = e^{\sqrt{3x+2}} \frac{d}{dx}[\sqrt{3x+2}]$

$$= e^{\sqrt{3x+2}} \; \frac{1}{2}(3x+2)^{-1/2} \frac{d}{dx}[3x+2]$$

$$= e^{\sqrt{3x+2}} \; \frac{1}{2} \cdot \frac{1}{\sqrt{3x+2}} \cdot 3$$

$$= \frac{3}{2\sqrt{3x+2}} e^{\sqrt{3x+2}}$$

9. $\frac{d}{dx}[3x^2 + e^{-x^3} + 4e^{x^2}] = \frac{d}{dx}[3x^2] + \frac{d}{dx}[e^{-x^3}] + \frac{d}{dx}[4e^{x^2}]$

$$= 6x + e^{-x^3} \frac{d}{dx}[-x^3] + 4e^{x^2} \frac{d}{dx}[x^2]$$

$$= 6x + e^{-x^3}(-3x^2) + 4e^{x^2}(2x)$$

$$= 6x - 3x^2 e^{-x^3} + 8xe^{x^2}$$

13. Using the product rule, we have

$$\frac{d}{dx}[x^3 e^{4x}] = x^3 \cdot \frac{d}{dx}[e^{4x}] + e^{4x} \cdot \frac{d}{dx}[x^3]$$

$$= x^3 \cdot e^{4x} \cdot \frac{d}{dx}[4x] + e^{4x} \cdot (3x^2)$$

$$= 4x^3 e^{4x} + 3x^2 e^{4x}$$

$$= x^2 e^{4x}(4x+3)$$

17. Using the quotient rule when $y = (e^x - e^{-x})/(e^x + e^{-x})$, we have

$$\frac{dy}{dx} = \frac{(e^x + e^{-x})\frac{d}{dx}[e^x - e^{-x}] - (e^x - e^{-x})\frac{d}{dx}[e^x + e^{-x}]}{(e^x + e^{-x})^2}$$

$$= \frac{(e^x + e^{-x})(e^x + e^{-x}) - (e^x - e^{-x})(e^x - e^{-x})}{(e^x - e^{-x})^2}$$

$$= \frac{(e^x + e^{-x})^2 - (e^x - e^{-x})^2}{(e^x + e^{-x})^2}$$

Recall that $a^2 - b^2 = (a+b)(a-b)$. We apply this result in the numerator with

$$a = e^x + e^{-x} \quad \text{and} \quad b = e^x - e^{-x}$$

so that

$$a + b = 2e^x$$

and

$$a - b = 2e^{-x}$$

Thus,

$$\frac{dy}{dx} = \frac{(2e^x)(2e^{-x})}{(e^x + e^{-x})^2} = \frac{4}{(e^x + e^{-x})^2}$$

21. $\dfrac{dy}{dx} = \dfrac{1}{2x^2 + 3x - 1} \dfrac{d}{dx}[2x^2 + 3x - 1]$

$= \dfrac{1}{2x^2 + 3x - 1} [\dfrac{d}{dx}(2x^2) + \dfrac{d}{dx}(3x) - \dfrac{d}{dx}(1)]$

$= \dfrac{1}{2x^2 + 3x - 1}(4x + 3) = \dfrac{4x + 3}{2x^2 + 3x - 1}$

23. Using the power rule, we have

$$\frac{d}{dx}[(\ln x)^5] = 5 (\ln x)^4 \frac{d}{dx}[\ln x]$$

$$= 5 (\ln x)^4 \cdot \frac{1}{x}$$

$$= \frac{5 (\ln x)^4}{x}$$

33. Using the product rule when $y = x^2 \ln x$, we have

$$\frac{dy}{dx} = x^2 \frac{d}{dx}[\ln x] + \ln x \frac{d}{dx}[x^2]$$

$$= (x^2)\frac{1}{x} + (\ln x)(2x)$$

$$= x + 2x \ln x$$

$$= x(1 + 2 \ln x)$$

37. When $y = \ln(e^x + x)$, we have

$$\frac{dy}{dx} = \frac{1}{e^x + x} \frac{d}{dx}[e^x + x]$$

$$= \frac{1}{e^x + x}(e^x + 1) = \frac{e^x + 1}{e^x + x}$$

43. Using (8), we have

$$\frac{d}{dx}[4^{-3x}] = 4^{-3x} \ln 4 \frac{d}{dx}[-3x]$$

$$= 4^{-3x} \ln 4 \,[-3]$$

$$= -3(4^{-3x} \ln 4)$$

55. $\frac{d}{dx}[x^3 \log x] = x^3 \frac{d}{dx}[\log x] + \log x \frac{d}{dx}[x^3]$

Using (17), we have

$$\frac{d}{dx} x^3 \log x = x^3 \left[\frac{1}{x \ln 10}\right] + \log x \,[3x^2]$$

$$= \frac{x^2}{\ln 10} + 3x^2 \log x$$

$$= x^2 \left[3 \log x + \frac{1}{\ln 10}\right]$$

59. When $y = x + e^x$, $dy/dx = 1 + e^x$. Thus, when $x = 0$ the slope is

$$\left.\frac{dy}{dx}\right|_{x=0} = 1 + e^0 = 1 + 1 = 2$$

65. When $y = xe^x$, we have

$$\frac{dy}{dx} = x \frac{d}{dx}[e^x] + e^x \frac{d}{dx}[x] = xe^x + e^x(1)$$

$$= e^x(x + 1)$$

and

5: The Exponential and Logarithmic Functions 123

$$\frac{d^2y}{dx^2} = e^x \frac{d}{dx}[x+1] + (x+1)\frac{d}{dx}[e^x]$$

$$= e^x(1) + (x+1)e^x$$

$$= e^x(x+2)$$

Now, $e^x > 0$ for all x so that dy/dx is zero only at x = -1. Since dy/dx exists for all values of x, x = -1 is the only critical number. We analyze the sign of dy/dx in the figure below.

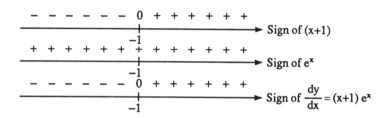

Thus, y is increasing when x > -1 and y is decreasing when x < -1. A relative minimum occurs at x = -1.

A similar analysis shows that $d^2y/dx^2 > 0$ for x > -2 and $d^2y/dx^2 < 0$ for x < -2. Thus, the graph is concave upward for x > -2 and concave downward for x < -2. An inflection point occurs when x = -2.

A sketch of $y = xe^x$ appears in the figure below.

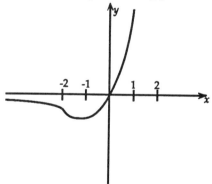

Exercise Set 5.4 (Page 320)

1. The exponential growth of quantity Q is given by

    $$Q = 5000e^{0.4t} \qquad (9)$$

    where t is measured in days.

(a) The amount present initially is

$$Q_0 = Q(0) = 5000$$

(b) The growth constant is 0.4, the coefficient of t.

(c) The differential equation of which (9) is a solution is

$$\frac{dQ}{dt} = kQ = 0.4Q$$

(d) When Q = 15,000, the rate at which Q is growing is

$$\frac{dQ}{dt} = (0.4)(15,000) = 6000$$

(e) At t = 5,

$$Q(5) = 5000e^{(0.4)(5)} = 36,945$$

3. (a) Since k = 0.05, the equation for exponential decay is

$$\frac{dQ}{dt} = -kQ = -0.05Q$$

(b) $Q = Q_0 e^{-kt}$ is the solution of the differential equation. Since $Q_0 = 8000$ and k = 0.05, we have

$$Q = 8000e^{-0.05t}$$

(c) At t = 30, the amount Q remaining is

$$Q = 8000e^{(-0.05)(30)} = 1785$$

Thus after 30 years

$$8000 - 1785 = 6215 \text{ pounds of the quantity has decayed.}$$

21. Since a fossil retains 4/5 of its original carbon-14 content (k = 0.00012), we have

$$\frac{4}{5}Q_0 = Q_0 e^{-0.00012t}$$

$$0.8 = e^{-0.00012t}$$

$$\ln 0.8 = -0.00012t$$

or

$$t = \frac{-\ln 0.8}{0.00012} = 1859.53$$

Thus, the fossil is approximately 1860 years old.

23. (a) We measure time in years since 1950 and Q in billions of barrels. Then $Q_0 = 1$ and when $t = 27$ (1977 - 1950) we have $Q = 2.6$. Thus

$$2.6 = (1)e^{k(27)} = e^{27k}$$

$$\ln 2.6 = 27k$$

$$k = \frac{\ln 2.6}{27} = 0.0354$$

(b) In the year 2000,

$$t = 2000 - 1950 = 50$$

and

$$Q = e^{50k} = e^{50 \ln(2.6)/27} = 5.8677$$

Thus, the oil consumption for the year 2000 is approximately 5.87 billion barrels per year.

25. The half-life of strontium-90 is 28 years. Thus

$$28 = \frac{\ln 2}{k}$$

or

$$k = \frac{\ln 2}{28} = 0.0248$$

is the decay constant.

If 90% of the radiation has disappeared, only 10% of the initial amount Q_0 remains at the time t when the area will be fit once more for human habitation. Thus we have

$$Q = Q_0 e^{-kt}$$

$$.10 Q_0 = Q_0 e^{-0.0248t}$$

$$.10 = e^{-0.0248t}$$

$$\frac{\ln .10}{-0.0248} = t$$

$$92.85 = t$$

It will be approximately 93 years before the area is fit for human habitation.

Review Exercises, (Page 324)

5. We are given $P = \$4000$, $r = 0.06$, $k = 4$, and $n = 6(4) = 24$ conversions. Now

$$i = \frac{r}{k} = \frac{0.06}{4} = 0.015$$

and

$$S = (P + i)^n = 4000(1.015)^{24} = 5718$$

Thus, the future value is approximately $5718.00

9. We have $r = 0.09$. The effective rate of interest is

$$V = e^r - 1 = e^{0.09} - 1 = 0.0942$$

The effective rate is 9.42%.

19. $\frac{d}{dx}[e^{x\sqrt{2x+1}}] = e^{x\sqrt{2x+1}} \frac{d}{dx}[x\sqrt{2x+1}]$

$$= e^{x\sqrt{2x+1}} (x \frac{d}{dx}[(2x+1)^{1/2}] + \sqrt{2x+1} \frac{d}{dx}[x])$$

$$= e^{x\sqrt{2x+1}} (x \cdot \frac{1}{2}(2x+1)^{-1/2} \frac{d}{dx}[2x+1] + \sqrt{2x+1}\,(1))$$

5: The Exponential and Logarithmic Functions 127

$$= e^{x\sqrt{2x+1}}(x \cdot \frac{1}{2} \cdot \frac{1}{\sqrt{2x+1}}(2) + \sqrt{2x+1})$$

$$= e^{x\sqrt{2x+1}}(\frac{x}{\sqrt{2x+1}} + \sqrt{2x+1})$$

$$= e^{x\sqrt{2x+1}}(\frac{x}{\sqrt{2x+1}} + \frac{2x+1}{\sqrt{2x+1}})$$

$$= \frac{3x+1}{\sqrt{2x+1}} e^{x\sqrt{2x+1}}$$

25. Now

$$y = \ln\sqrt[3]{x^2} = \ln x^{2/3} = \frac{2}{3} \ln x$$

so

$$\frac{dy}{dx} = \frac{2}{3} \cdot \frac{1}{x} = \frac{2}{3x}$$

29. $\frac{d}{dx}[3^{2x}] = 3^{2x} \ln 3 \frac{d}{dx}(2x) = 3^{2x} \ln 3 [2] = 2(3^{2x} \ln 3)$

31. $\frac{d}{dx} \log_2 5x = \frac{1}{5x \ln 2} \frac{d}{dx}[5x] = \frac{1}{5x \ln 2}(5) = \frac{1}{x \ln 2}$

33. When $y = x^2 + 5e^x$, $dy/dx = 2x + 5e^x$. Thus, the slope of the tangent line at $x = 0$ is

$$\left.\frac{dy}{dx}\right|_{x=0} = 2(0) + 5e^0 = 0 + 5 = 5$$

39. We are given $k = 0.133$

(a) The appropriate differential equation is

$$\frac{dQ}{dt} = -kQ = -0.133Q$$

Its solution is $Q = Q_0 e^{-0.133t}$

(b) We want to find t when $Q = \frac{1}{2}Q_0$. We have

$$\frac{1}{2}Q_0 = Q_0 e^{-0.133t}$$

$$\frac{1}{2} = e^{-0.133t}$$

$$\ln \frac{1}{2} = -0.133t$$

$$-\ln 2 = -0.133t$$

$$t = \frac{\ln 2}{0.133} = 5.21 \text{ years}$$

(c) We want to find t when $Q = \frac{1}{4}Q_0$. We have

$$\frac{1}{4}Q_0 = Q_0 e^{-0.133t}$$

$$\frac{1}{4} = e^{-0.133t}$$

$$\ln \frac{1}{4} = -0.133t$$

$$-\ln 4 = -0.133t$$

$$t = \frac{\ln 4}{0.133} = 10.42 \text{ years}$$

Chapter Test, (Page 325)

1.

x	-2	-1	0	1	2
$y = 3^x$	1/9	1/3	1	3	9

x	3	9	27
$y = \log_3 x$	1	2	3

5: *The Exponential and Logarithmic Functions* 129

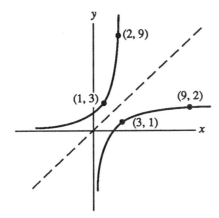

3. We are given P = 4000, r = 0.09, and S = 7,000. Thus we have

$$S = Pe^{rt}$$

$$7000 = 4000e^{0.09t}$$

or

$$e^{0.09t} = \frac{7000}{4000} = 1.75$$

$$0.9t = \ln 1.75$$

$$t = \frac{\ln 1.75}{.09} = 6.2 \text{ years}$$

5. Given $r = \log_b x$, $s = \log_b y$, and $t = \log_b z$. Then

$$\log_b \sqrt{\frac{xy^3}{z^4}} = \log_b \left(\frac{xy^3}{z^4}\right)^{1/2}$$

$$= \frac{1}{2}[\log_b x + \log_b y^3 - \log_b z^4]$$

$$= \frac{1}{2}\log_b x + \frac{3}{2}\log_b y - 2\log_b z$$

$$= \frac{1}{2}r + \frac{3}{2}s - 2t$$

7. (a) If $f(x) = 50e^{0.1x}$, then

$$f'(x) = 50e^{0.1x}\frac{d}{dx}(0.1x) = 50e^{0.1x}(0.1) = 5e^{0.1x}$$

(b) If $f(x) = \ln(4x^3 + x)$, then

$$f'(x) = \frac{1}{4x^3 + x} \frac{d}{dx}(4x^3 + x) = \frac{1}{4x^3 + x}(12x^2 + 1) = \frac{12x^2 + 1}{4x^3 + x}$$

(c) If $f(x) = \frac{x^4}{e^{3x}} = x^4 e^{-3x}$, then

$$f'(x) = x^4 \frac{d}{dx}(e^{-3x}) + e^{-3x}\frac{d}{dx}(x^4)$$

$$= x^4 e^{-3x}\frac{d}{dx}(-3x) + e^{-3x}(4x^3)$$

$$= x^4 e^{-3x}(-3) + 4x^3 e^{-3x}$$

$$= e^{-3x}(4x^3 - 3x^4)$$

$$= \frac{4x^3 - 3x^4}{e^{3x}}$$

(d) If $f(x) = e^{\sqrt{x}} = e^{x^{1/2}}$, then

$$f'(x) = e^{x^{1/2}} \frac{d}{dx}(x^{1/2})$$

$$= e^{x^{1/2}}(\frac{1}{2}x^{-1/2})$$

$$= \frac{e^{\sqrt{x}}}{2\sqrt{x}}$$

9. (a) Since $k = 0.03$, we have

$$\frac{dQ}{dt} = 0.03\, Q$$

where Q is the population at time t. The solution is

$$Q = Q_0 e^{0.03t}$$

Since $Q_0 = Q(0) = 4000$, we have

$$Q = 4000 e^{0.03t}$$

(b) At $t = 10$, we have

$$Q = 4000e^{(0.03)(10)} = 5399$$

(c) When the population doubles, $Q = 8000$ and we have

$$8000 = 4000e^{0.03t}$$

$$e^{0.03t} = 2$$

$$t = \frac{\ln 2}{0.03} = 23.1 \text{ hours}$$

6 Antidifferentiation

Key Ideas for Review

* $F(x)$ is an antiderivative of $f(x)$ if $F'(x) = f(x)$ for all x in the domain of f.

* Any two antiderivatives of the same function differ by a constant amount.

* If $F(x)$ is any antiderivative of $f(x)$, then every antiderivative of $f(x)$ is given by

$$\int f(x)\, dx) = F(x) + C$$

  where C is an arbitrary constant.

* $\int k\, dx = kx + C$

* $\int x^r\, dx = \dfrac{x^{r+1}}{r+1} + C, \quad r \neq -1$

* $\int \dfrac{1}{x}\, dx = \ln |x| + C, \quad x \neq 0$

* $\int e^x\, dx = e^x + C$

* $\int a^x\, dx = \dfrac{a^x}{\ln a} + C$

* $\int kf(x)\, dx = k \int f(x)\, dx$ (Beware: only constants "factor out" of an integral.)

* $\int [f(x) + g(x)]\, dx = \int f(x)\, dx + \int g(x)\, dx$

* $\int [f(x) - g(x)]\, dx = \int f(x)\, dx - \int g(x)\, dx$

* $\int [f(x)]^k f'(x)\, dx =$
$\dfrac{1}{k+1} [f(x)]^{k+1} + C, \quad k \neq -1$

or

$$\int u^k \, du = \frac{1}{k+1} u^{k+1} + C$$

where $u = f(x)$, $k \neq -1$

* $\int e^{f(x)} f'(x) \, dx = e^{f(x)} + C$

or

$\int e^u \, du = e^u + C$, where $u = f(x)$

* $\int \frac{f'(x)}{f(x)} \, dx = \ln |f(x)| + C$, $f(x) \neq 0$

or

$\int \frac{1}{u} \, du = \ln |u| + C$, where $u = f(x) \neq 0$

* Integration by substitution:

    If $\quad \int f(x) \, dx = F(x) + C$

    then $\int f(ax + b) = \frac{1}{a} F(ax + b) + C$

* Integration by parts:

    $\int u(x) \, v'(x) \, dx = u(x) \, v(x) - \int v(x) \, u'(x) \, dx$

    or

    $\int u \, dv = uv - \int v \, du$

* Before integration tables can be used, it may be necessary to algebraically manipulate the integrand.

* An expression that yields all the solutions of a first-order differential equation includes an arbitrary constant.

* To find the solution of an initial-value problem, first find all the solutions of the associated differential equation. Then determine a value for the constant in this expression so that the result also satisfies the associated initial condition.

*   A separable first-order differential equation is one that can be written as $dy/dx = f(x) g(y)$, where $f(x)$ is a function of x only and $g(y)$ is a function of y only. Solve these equations by separating the variables and integrating both sides of the resulting equation.

*   The differential equation $dQ/dt = kQ$ describes unrestricted (exponential) growth. The expression $Q(t) = Ce^{kt}$ yields all the solutions. Moreover, $C = Q_0$, where $Q_0 = Q(0)$.

*   The differential equation $dQ/dt = k(L - Q)$ describes simple restricted growth. The expression $Q(t) = L - Ce^{-Lkt}$ yields all the solutions. Moreover, $C = L - Q_0$, where $Q_0 = Q(0)$.

*   The differential equation $dP/dt = kP(L - P)$ describes logistic growth. It behaves like unrestricted growth for small values of P and like restricted growth as P approaches L. The expression $P(t) = L/(1 + Ce^{-Lkt})$, called the logistic equation, together with the constant function $P(t) = 0$ yields all the solutions.

Exercise Set 6.1 (Page 337)

3. $\int s^6 \, ds = \dfrac{s^7}{7} + C$

11. $\int 3e^u \, du = 3e^u + C$

13. $\int \dfrac{10}{x} \, dx = 10 \ln |x| + C$

15. $\int 3\sqrt{x} \, dx = \int x^{1/3} \, dx = \dfrac{x^{4/3}}{4/3} + C = \dfrac{3}{4} x^{4/3} + C$

23. $\int (2t^{2/3} - 3e^t + \dfrac{2}{t}) \, dt = \dfrac{2t^{5/3}}{5/3} - 3e^t + 2 \ln |t| + C$

    $\qquad = \dfrac{6}{5} t^{5/3} - 3e^t + 2 \ln |t| + C$

29. $F(x)$ is an antiderivative of $f(x)$ if and only if $F'(x) = f(x)$. $F(x) = e^{3x}$ is not an antiderivative of $f(x) = e^{3x}$ since $F'(x) = 3e^{3x} \neq f(x)$.

35. $\int f(t) \, dt = F(t) + C$ if and only if $F'(t) = f(t)$. Thus

    $\int e^{5t} \, dt \neq e^{5t} + C$

since

$$F'(t) = \frac{d}{dt}[e^{5t}] = 5e^{5t} \neq e^{5t} = f(t)$$

37. $\int \frac{x^3 + 2x^2}{3x^5} dx = \int [\frac{x^3}{3x^5} + \frac{2x^2}{3x^5}] dx$

$= \frac{1}{3}\int [x^{-2} + 2x^{-3}] dx$

$= \frac{1}{3}[\frac{x^{-1}}{-1} + \frac{2x^{-2}}{-2}] + C$

$= \frac{-1}{3x} - \frac{1}{3x^2} + C$

41. $\int \sqrt{x}(x+3) dx = \int x^{1/2}(x+3) dx$

$= \int (x^{3/2} + 3x^{1/2}) dx$

$= \frac{2}{5}x^{5/2} + 3(\frac{2}{3})x^{3/2} + C$

$= \frac{2}{5}x^{5/2} + 2x^{3/2} + C$

43. Since $C'(x) = 3\sqrt{x} + 2x + 5$, we find $C(x)$ by computing the indefinite integral of $C'(x)$. Hence,

$C(x) = \int C'(x) dx$

$= \int (3x^{1/2} + 2x + 5) dx$

$= 3(x^{3/2}/(3/2)) + 2(x^2/2) + 5x + k$

$= 2x^{3/2} + x^2 + 5x + k$

The fixed cost is the cost when $x = 0$. Then,

$$C(0) = 2000 = 2(0)^{3/2} + (0)^2 + 5(0) + k$$

so

$$k = 2000$$

Hence, the desired cost function is

136 Study Guide

$$C(x) = 2x^{3/2} + x^2 + 5x + 2000$$

47. Let P(t) be the city's population (in thousands) t years after incorporation. Then, P(t) is an antiderivative of $P'(t) = 10 + 6t^{1/5}$. Hence,

$$P(t) = \int P'(t)\, dt$$

$$= \int (10 + 6t^{1/5})\, dt$$

$$= 10t + \frac{6t^{6/5}}{6/5} + C = 10t + 5t^{6/5} + C$$

The population at incorporation is 20,000 so that $P(0) = 20$. We use this condition to determine the constant of integration. Now,

$$P(0) = 10(0) + 5(0)^{6/5} + C = 20$$

So

$$C = 20$$

and

$$P(t) = 10t + 5t^{6/5} + 20$$

The population 32 years after incorporation is determined by

$$P(32) = 10(32) + 5(32)^{6/5} + 20 = 660$$

Thus, there are 660,000 people in the city 32 years after its incorporation.

Exercise Set 6.2, (Page 347)

5. Let $u = 1 - w^3$ so that $du = -3w^2\, dw$. Then $w^2\, dw = -\frac{1}{3} du$ and

$$\int w^2 (1-w^3)^7\, dw = \int u^7 \left(-\frac{1}{3} du\right) = -\frac{1}{3}\int u^7\, du$$

$$= -\frac{1}{3}\frac{u^8}{8} + C = -\frac{1}{3}\frac{(1-w^3)^8}{8} + C$$

$$= -\frac{1}{24}(1-w^3)^8 + C$$

7. Let $u = x^2 + 9$ so that $du = 2x\, dx$. Then

$$x\, dx = \frac{1}{2} du$$

and

$$\int \frac{x}{(x^2+9)^3}\, dx = \int \frac{1}{u^3}\left(\frac{1}{2} du\right) = \frac{1}{2}\int u^{-3}\, du$$

$$= \frac{1}{2} \cdot \frac{u^{-2}}{-2} + C$$

$$= -\frac{1}{4u^2} + C = -\frac{1}{4(x^2+9)^2} + C$$

13. Let $u = x^3 + 2$ so that $du = 3x^2\, dx$. Then

$$\int (x^3+2)^{1/2} 3x^2\, dx = \int u^{1/2}\, du = \frac{2}{3} u^{3/2} + C$$

$$= \frac{2}{3}(x^3+2)^{3/2} + C$$

15. Let $u = t^4 + 5$ so that $du = 4t^3\, dt$. Then

$$t^3\, dt = \frac{1}{4} du$$

and

$$\int \sqrt{t^4+5}\, t^3\, dt = \int \sqrt{u}\, \frac{1}{4} du = \frac{1}{4}\int u^{1/2}\, du$$

$$= \frac{1}{4} \cdot \frac{u^{3/2}}{3/2} + C = \frac{1}{6} u^{3/2} + C$$

$$= \frac{1}{6}(t^4+5)^{3/2} + C$$

21. Let $u = x^2$ so that $du = 2x\, dx$. Then

$$\int 2xe^{x^2}\,dx = \int e^u\,du = e^u + C = e^{x^2} + C$$

27. Let $u = 1 + e^y$ so that $du = e^y\,dy$. Then

$$\int e^y(1+e^y)\,dy = \int u\,du = \frac{u^2}{2} + C = \frac{1}{2}(1+e^y)^2 + C$$

31. Let $u = 2 - x$ so that $du = -dx$. Then

$$dx = -du$$

and

$$\int \frac{1}{2-x}\,dx = \int \frac{1}{u}(-du) = -\int \frac{1}{u}\,du = -\ln|u| + C$$

$$= -\ln|2-x| + C$$

33. Let $u = 3s^2 + 2$ so that $du = 6s\,ds$. Then $s\,ds = \frac{1}{6}du$ and

$$\int \frac{s}{3s^2+2}\,ds = \int \frac{1/6\,du}{u} = \frac{1}{6}\int \frac{du}{u} = \frac{1}{6}\ln|u| + C$$

$$= \frac{1}{6}\ln|3s^2+2| + C$$

37. Let $u = \ln x$ so that $du = (1/x)\,dx$. Then

$$\int \frac{(\ln x^2)}{x}\,dx = \int u^2\,du = \frac{u^3}{3} + C$$

$$= \frac{(\ln x)^3}{3} + C$$

41. Let $R(x)$ denote the revenue (in millions of dollars) from the manufacture and sale of x thousand units. Then, $R(x)$ is an antiderivative of $R'(x)$. Hence

$$R(x) = \int R'(x)\,dx = \int \frac{x}{\sqrt{x^2+16}}\,dx$$

Let $u = x^2 + 16$ so that $du = 2x\,dx$. Then,

$$x \, dx = \frac{1}{2} du$$

and

$$R(x) = \int \frac{x}{\sqrt{x^2 + 16}} \, dx = \int \frac{1/2}{\sqrt{u}} \, du = \frac{1}{2} \int u^{-1/2} \, du$$

$$= \frac{1}{2} \cdot \frac{u^{1/2}}{1/2} + C = \sqrt{u} + C$$

or

$$R(x) = \sqrt{x^2 + 16} + C$$

The revenue is zero when the level of production is zero, that is,

$$R(0) = 0$$

We use this condition to determine the constant of integration.

$$R(0) = \sqrt{0^2 + 16} + C = 0$$

$$\sqrt{16} + C = 0$$

$$C = -4$$

and

$$R(x) = \sqrt{x^2 + 16} - 4$$

Now

$$R(3) = \sqrt{(3)^2 + 16} - 4 = \sqrt{25} - 4 = 1$$

Thus, the revenue is 1 million dollars when the level of production is 3000 units.

Exercise Set 6.3, (Page 355)

1. Let

$$u(x) = x \text{ and } v'(x) = 3e^{3x}$$

so that

$u'(x) = 1$ and $v(x) = \int v'(x)\, dx = \int 3e^{3x}\, dx = e^{3x}$

Then

$$\int 3xe^{3x}\, dx = \int u(x)\, v'(x)\, dx$$
$$= u(x)\, v(x) - \int v(x)\, u'(x)\, dx$$
$$= xe^{3x} - \int e^{3x} \cdot 1\, dx$$
$$= xe^{3x} - \int e^{3x}\, dx$$
$$= xe^{3x} - \frac{1}{3}e^{3x} + C$$

5. Let

$$u(t) = \ln t \text{ and } v'(t) = \sqrt{t} = t^{1/2}$$

so that

$$u'(t) = \frac{1}{t} \text{ and } v(t) = \int v'(t)\, dt = \int t^{1/2}\, dt = \frac{2}{3}t^{3/2}$$

Then

$$\int \sqrt{t}\, \ln t\, dt = \int u(t)\, v'(t)\, dt$$
$$= \frac{2}{3}t^{3/2} \ln t - \int \frac{2}{3}t^{3/2} \frac{1}{t}\, dt$$
$$= \frac{2}{3}t^{3/2} \ln t - \frac{2}{3}\int t^{1/2}\, dt$$
$$= \frac{2}{3}t^{3/2} \ln t - \frac{2}{3} \cdot \frac{2}{3}t^{3/2} + C$$
$$= \frac{2}{3}t^{3/2} \ln t - \frac{4}{9}t^{3/2} + C$$

7. Let

$$u(x) = (\ln x)^2 \text{ and } v'(x) = x$$

so that

$$u'(x) = \frac{2 \ln x}{x} \text{ and } v(x) = \int v'(x) \, dx = \int x \, dx = \frac{x^2}{2}$$

Then

$$\int x (\ln x)^2 \, dx = \int u(x) v'(x) \, dx$$

$$= u(x) v(x) - \int v(x) u'(x) \, dx$$

$$= (\ln x)^2 \left(\frac{x^2}{2}\right) - \int \frac{x^2}{2} \left(\frac{2 \ln x}{x}\right) dx$$

$$= \frac{x^2}{2} (\ln x)^2 - \int x \ln x \, dx$$

We now use integration by parts again. Let $u = \ln x$ and $v'(x) = x$ so that $u'(x) = \frac{1}{x}$ and $v(x) = \int v'(x) \, dx = \int x \, dx = \frac{x^2}{2}$. Then

$$\int x \ln x \, dx = \int u(x) v'(x) \, dx$$

$$= u(x) v(x) - \int v(x) u'(x) \, dx$$

$$= (\ln x) \frac{x^2}{2} - \int \frac{x^2}{2} \left(\frac{1}{x}\right) dx$$

$$= \frac{x^2}{2} \ln x - \int \frac{1}{2} x \, dx$$

$$= \frac{x^2}{2} \ln x - \frac{1}{2} \cdot \frac{x^2}{2} + C$$

Thus

$$\int x (\ln x)^2 \, dx = \frac{x^2}{2} (\ln x)^2 - \left[\frac{x^2}{2} \ln x - \frac{1}{4} x^2 + C\right]$$

$$= \frac{x^2}{2} (\ln x)^2 - \frac{x^2}{2} \ln x + \frac{1}{4} x^2 + C$$

11. Let

$$u(x) = x \text{ and } v'(x) = \sqrt{x+3} = (x+3)^{1/2}$$

so that

$$u'(x) = 1 \text{ and } v(x) = \int v'(x)\,dx = \int (x+3)^{1/2}\,dx$$

It can be shown using the method of substitution that

$$v(x) = \frac{2}{3}(x+3)^{3/2}$$

Then

$$\int x\sqrt{x+3}\,dx = \int u(x)\,v'(x)\,dx$$

$$= u(x)\,v(x) - \int v(x)\,u'(x)\,dx$$

$$= x \cdot \frac{2}{3}(x+3)^{3/2} - \int \frac{2}{3}(x+3)^{3/2} \cdot 1\,dx$$

$$= \frac{2x}{3}(x+3)^{3/2} - \frac{2}{3}\int (x+3)^{3/2}\,dx$$

Again the method of substitution is required. The result is

$$\int (x+3)^{3/2}\,dx = \frac{2}{5}(x+3)^{5/2} + C_1$$

Thus

$$\int x\sqrt{x+3}\,dx = \frac{2x}{3}(x+3)^{3/2} - \frac{4}{15}(x+3)^{5/2} + C$$

15. Let

$$u(x) = \ln(x+1) \text{ and } v'(x) = 1$$

so that

$$u'(x) = \frac{1}{x+1} \text{ and } v(x) = \int v'(x)\,dx = \int 1\,dx = x$$

then

$$\int \ln(x+1)\,dx = \int v(x)\,u'(x)\,dx$$

$$= u(x)\,v(x) - \int v(x)\,u'(x)\,dx$$

$$= [\ln (x + 1)] (x) - \int x \left(\frac{1}{x+1}\right) dx$$

We use substitution to find $\int \frac{x}{x+1} dx$. Let $t = x + 1$ so that $x = t - 1$ and $dx = dt$. Then

$$\int \frac{x}{x+1} dx = \int \frac{t-1}{t} dt = \int \left(1 - \frac{1}{t}\right) dt$$

$$= t - \ln |t| + C_1 = (x + 1) - \ln |x + 1| + C_1$$

Thus

$$\int \ln (x + 1) dx = x \ln (x + 1) - [(x + 1) - \ln |x + 1| + C_1]$$

$$= (x + 1) \ln (x + 1) - x + C$$

19. Let

$$u = \ln x \text{ and } v'(x) = x^3$$

so that

$$u'(x) = \frac{1}{x} \text{ and } v(x) = \int v'(x) \, dx = \int x^3 \, dx = \frac{x^4}{4}$$

Then

$$\int x^3 \ln x \, dx = \int u(x) v'(x) \, dx$$

$$= u(x) v(x) - \int v(x) u'(x) \, dx$$

$$= (\ln x) \left(\frac{x^4}{4}\right) - \int \left(\frac{x^4}{4}\right) \left(\frac{1}{x}\right) dx$$

$$= \frac{x^4 \ln x}{4} - \frac{1}{4} \int x^3 \, dx$$

$$= \frac{x^4}{4} \ln x - \frac{1}{4} \left(\frac{x^4}{4}\right) + C$$

$$= \frac{x^4}{4} \ln x - \frac{x^4}{16} + C$$

27. Let R(x) denote the revenue (in millions of dollars) from the manufacture and sale of x thousand units. Then

$$R(x) = \int R'(x)\, dx = \int 4x - xe^{-0.2x}\, dx$$

$$= 2x^2 - \int xe^{-0.2x}\, dx$$

We determine $\int xe^{-0.2x}\, dx$ by parts. Let

$$u(x) = x \text{ and } v'(x) = e^{-0.2x}$$

so that

$$u'(x) = 1 \text{ and } v(x) = \int v'(x)\, dx = \int e^{-0.2x}\, dx = -5e^{-0.2x}$$

The method of substitution was used above to determine v(x). Now

$$\int xe^{-0.2x}\, dx = \int u(x)\, v'(x)\, dx$$

$$= u(x)\, v(x) - \int v(x)\, u'(x)\, dx$$

$$= x(-5e^{-0.2x}) - \int (-5e^{-0.2x}) \cdot 1\, dx$$

$$= 5xe^{-0.2x} + 5\int e^{-0.2x}\, dx$$

$$= -5xe^{-0.2x} + 5v(x) + C_1$$

$$= -5xe^{-0.2x} - 25e^{-0.2x} + C_1$$

We substitute this into the expression for R(x) obtaining

$$R(x) = 2x^2 - (-5xe^{-0.2x} - 25e^{-0.2x}) + C_1$$

$$= 2x^2 + (5x + 25)e^{-0.2x} + C_1$$

Now the revenue is zero when the level of production is zero; that is, R(0) = 0. We use this condition to determine the constant of integration.

$$R(0) = 2(0)^2 + (5(0) + 25)e^{-0.2(0)} + C_1 = 0$$

so that

$$25 + C_1 = 0$$

or

$$C_1 = -25$$

and

$$R(x) = 2x^2 + (5x + 25)e^{-0.2x} - 25$$

Now

$$R(5) = 2(5)^2 + (5(5) + 25)e^{-0.2(5)} - 25$$

$$= 50 + 50e^{-1} - 25$$

$$= 25 + 50e^{-1}$$

Thus, when the level of production is 5000 units, the revenue is $25 + 50e^{-1}$ million dollars; that is, approximately \$43,393,972.

Exercise Set 6.4, (Page 363)

1. We use Formula 10

$$\int \frac{u}{au + b} du = \frac{u}{a} - \frac{b}{a^2} \ln|au + b| + C$$

with $u = x$, $a = 5$, and $b = 10$, to obtain

$$\int \frac{x}{5x + 10} dx = \frac{x}{5} - \frac{10}{(5)^2} \ln|5x + 10| + C$$

$$= \frac{x}{5} - \frac{2}{5} \ln|5x + 10| + C$$

5. We use Formula 28

$$\int \frac{1}{u^2 - a^2} du = \frac{1}{2a} \ln\left|\frac{u - a}{u + a}\right| + C$$

with $u = 7x$ so that $du = 7\,dx$ and $a = 2$. We have

$$\int \frac{1}{49x^2 - 4} dx = \int \frac{1/7\,du}{u^2 - a^2} = \frac{1}{7}[\frac{1}{2(2)} \ln\left|\frac{7x - 2}{7x + 2}\right|] + C$$

$$= \frac{1}{28} \ln \left| \frac{7x-2}{7x+2} \right| + C$$

11. We use Formula 39

$$\int e^{au} \, du = \frac{1}{a} e^{au} + C$$

with $u = x$ and $a = 5/2$ to obtain

$$\int e^{5x/2} \, dx = \frac{1}{5/2} e^{5x/2} + C$$

$$= \frac{2}{5} e^{5x/2} + C$$

13. We use Formula 4

$$\int u^n \, du = \frac{u^{n+1}}{n+1} + C$$

with $u = 4t + 3$ so that $du = 4dt$ and $n = -3$. We have

$$\int \frac{dt}{(4t+3)^3} = \int u^{-3} \frac{1}{4} \, du = \frac{1}{4} \frac{u^{-3+1}}{-3+1} + C$$

$$= \frac{1}{4} \frac{(4t+3)^{-2}}{-2} + C = \frac{-1}{8(4t+3)^2} + C$$

23. We use Formula 21

$$\int u^2 \sqrt{au+b} \, du = \frac{2}{105a^3} (15a^2u^2 - 12abu + 8b^2)(au+b)^{3/2} + C$$

with $u = x$, $a = 5$, and $b = 8$ to obtain

$$\int x^2 \sqrt{5x+8} \, dx = \frac{2}{105(5)^3} (15(5)^2 x^2 - 12(5)(8)x + 8(8)^2)(5x+8)^{3/2} + C$$

$$= \frac{2}{13125} (375x^2 - 480x + 512)(5x+8)^{3/2} + C$$

29. We use Formula 26

$$\int \frac{du}{u\sqrt{au+b}} = \frac{1}{\sqrt{b}} \ln \left| \frac{\sqrt{au+b}-\sqrt{b}}{\sqrt{au+b}+\sqrt{b}} \right| + C \text{ where } b > 0$$

with $u = t$, $a = 1$, and $b = 4$ to obtain

$$\int \frac{dt}{t\sqrt{t+4}} = \frac{1}{\sqrt{4}} \ln \left| \frac{\sqrt{t+4}-\sqrt{4}}{\sqrt{t+4}+\sqrt{4}} \right| + C$$

$$= \frac{1}{2} \ln \left| \frac{\sqrt{t+4}-2}{\sqrt{t+4}+2} \right| + C$$

33. We use Formula 42

$$\int u^n e^{ku} du = \frac{u^n e^{ku}}{k} - \frac{n}{k} \int u^{n-1} e^{ku} du \qquad k \neq 0$$

with $u = t$, $n = 2$, and $k = 5$, to obtain

$$\int t^2 e^{5t} dt = \frac{t^2}{5} e^{5t} - \frac{2}{5} \int t e^{5t} dt$$

We now use Formula 40 on the last integral

$$\int t e^{5t} dt = \frac{e^{5t}}{(5)^2} (5t - 1) + C$$

Thus

$$\int t^2 e^{5t} dt = \frac{t^2}{5} e^{5t} - \frac{2}{5} \left[\frac{e^{5t}}{25} (5t - 1) \right] + C$$

$$= \frac{t^2}{5} e^{5t} - \frac{2}{25} t e^{5t} + \frac{2}{125} e^{5t} + C$$

Exercise Set 6.5, (Page 371)

3. If $y = x^2 - 1$, then $y' = 2x$. Substituting these expressions into the given equation, we obtain

$$(y')^2 - 4y = (2x)^2 - 4(x^2 - 1) = 4x^2 - 4x^2 + 4 = 4$$

Thus, $y = x^2 - 1$ is a solution of $(y')^2 - 4y = 4$.

19. We write the given equation as

$$\frac{dy}{dx} = 3x^2 y$$

Assuming y is not the zero function, separation of variables yields

$$\frac{dy}{y} = 3x^2 \, dx$$

Integrating both sides, we obtain

$$\int \frac{dy}{y} = \int 3x^2 \, dx$$

$$\ln |y| = x^3 + C_1$$

$$|y| = e^{x^3 + C_1}$$

or

$$y = \pm e^{C_1} e^{x^3}$$

Since C_1 is an arbitrary constant, so are $\pm e^{C_1}$. Thus, we have shown every nonzero solution is of the form $y = Ce^{x^3}$ where C is an arbitrary constant. The constant solution of $y = 0$ is also of this form (pick $C = 0$). Thus all solutions are of the form $y = Ce^{x^3}$.

21. We write the given equation as

$$\frac{dy}{dx} = 5 + y$$

Separation of variables yields

$$\frac{dy}{5+y} = dx$$

Integrating both sides, we obtain

$$\int \frac{dy}{5+y} = \int dx$$

$$\ln |5+y| = x + C$$

$$|5+y| = e^{x + C_1}$$

6: Antidifferentiation 149

$$5 + y = \pm e^{C_1} e^x$$

or

$$y = \pm e^{C_1} e^x - 5$$

Since C_1 is an arbitrary constant, so are $\pm e^{C_1}$. Thus every solution is on the form $Ce^x - 5$, where C is an arbitrary constant.

27. We write the differential equation as

$$2x\frac{dy}{dx} = x^2 + 1$$

Separation of variables yields

$$dy = \frac{x^2 + 1}{2x} dx$$

Integrating both sides, we obtain

$$\int dy = \int \frac{x^2 + 1}{2x} dx$$

$$y = \int (x + \frac{1}{x}) dx$$

$$= \frac{1}{2} \int (x + \frac{1}{x}) dx$$

$$= \frac{1}{4} x^2 + \frac{1}{2} \ln |x| + C$$

Thus, every solution of the differential equation is of the form

$$y = \frac{1}{4} x^2 + \frac{1}{2} \ln |x| + C$$

31. We write the differential equation as

$$(4 + x^2) \frac{dy}{dx} = x$$

Separation of variables yields

$$dy = \frac{x}{4 + x^2} dx$$

Integrating both sides, we obtain

$$\int dy = \int \frac{x}{4+x^2} dx$$

or

$$y = \int \frac{x}{4+x^2} dx$$

Let $u = 4 + x^2$ so that $du = 2x\, dx$ and $x\, dx = \frac{1}{2} du$. Then

$$y = \int \frac{1}{u}(\frac{1}{2} du) = \frac{1}{2} \int \frac{1}{u} du$$

$$= \frac{1}{2} \ln |u| + C$$

$$= \frac{1}{2} \ln |4 + x^2| + C$$

$$= \frac{1}{2} \ln (4 + x^2) + C$$

since $4 + x^2 > 0$ for all x. Thus, every solution of the differential equation is of the form

$$y = \frac{1}{2} \ln(4 + x^2) + C$$

33. We write the differential equation as

$$\frac{dy}{dx} = -5x$$

Separating variables yields

$$dy = -5x\, dx$$

We integrate both sides to obtain

$$\int dy = \int -5x\, dx$$

$$y = -\frac{5}{2}x^2 + C$$

Substituting y = 2 when x = 0, gives

$$2 + 0 = C \quad \text{or} \quad C = 2$$

Thus

$$y = -\frac{5}{2}x^2 + 2$$

solves the initial value problem.

39. We wrtie the differential equation as

$$(x^2 + 2)\frac{dy}{dx} = x$$

Separating variables yields

$$dy = \frac{x}{x^2 + 2}dx$$

We integrate both sides to obtain

$$\int dy = \int \frac{x}{x^2 + 2}dx$$

$$y = \int \frac{x}{x^2 + 2}dx$$

Let $u = x^2 + 2$ so that $du = 2x\, dx$ or $x\, dx = \frac{1}{2}du$. Then

$$y = \int \frac{1}{u}(\frac{1}{2}du) = \frac{1}{2}\int \frac{1}{u}du$$

$$= \frac{1}{2}\ln |u| + C$$

$$= \frac{1}{2}\ln |x^2 + 2| + C$$

or

$$y = \frac{1}{2}\ln(x^2+2) + C$$

since $x^2 + 2 > 0$ for every x. Substituting $y = 3$ and $x = 2$, we obtain

$$3 = \frac{1}{2}\ln(2^2+2) + C$$

$$= \frac{1}{2}\ln 6 + C$$

or

$$C = 3 - \frac{1}{2}\ln 6$$

Thus,

$$y = \frac{1}{2}\ln(x^2+2) + 3 - \frac{1}{2}\ln 6$$

solves the initial-value problem.

Exercise Set 6.6, (Page 384)

3. Given $\frac{dP}{dt} = 10P(100 - P)$

   This differential equation is of the form

   $$\frac{dP}{dt} = kP(L - P)$$

   which expresses logistic growth. The solution is given by Theorem 2

   $$P = \frac{L}{1 + Ce^{-Lkt}}$$

   Since $k = 10$ and $L = 100$, we have

   $$P = \frac{100}{1 + Ce^{-100(10)t}} = \frac{100}{1 + Ce^{-1000t}}$$

6: Antidifferentiation   153

5. Given $\dfrac{dP}{dt} = 5(50 - P)$

This differential equation is of the form

$$\dfrac{dP}{dt} = k(L - P)$$

which expresses simple restricted growth. The solution is given by Theorem 1.

$$P = L - (L - P_0) e^{-kt}$$

Since $k = 5$, and $L = 50$, we have

$$P = 50 - (50 - P_0) e^{-5t}$$

Also, we are given $P = 12$ when $t = 0$. Thus

$$12 = 50 - (50 - P_0)$$

or

$$P_0 = 12.$$

The function

$$P = 50 - (50 - 12) e^{-5t} = 50 - 38 e^{-5t}$$

satisfies the initial-value problem.

9. (a) The differential equation

$$\dfrac{dQ}{dt} = k(200 - Q)$$

expresses simple restricted growth.

The solution is

$$Q = L - (L - Q_0) e^{-kt}$$

We have $L = 200$, so

$$Q = 200 - (200 - Q_0) e^{-kt}$$

We are given Q = 80 at t = 0. Hence

$$Q_0 = 80.$$

We are also given Q = 100 at t = 5. Thus

$$Q = 200 - (200 - 80) e^{-kt} = 200 - 120 e^{-kt}$$

$$100 = 200 - 120 e^{-5k}$$

$$-100 = -120 e^{-5k}$$

$$\frac{100}{120} = e^{-5k}$$

Taking the natural logarithm of both sides, we have

$$-5k = \ln \frac{5}{6}$$

$$k = -\frac{1}{5} \ln \frac{5}{6} = 0.036$$

Thus, on the t'th day, the number of units of work produced will be

$$Q = 200 - 120 e^{-0.036t}$$

On the 15th day,

$$Q = 200 - 120 e^{-0.036(15)} = 130.07$$

Thus, approximately 130 units of work will be produced on the 15th day.

On the 30th day

$$Q = 200 - 120 e^{-0.036(30)} = 159.24$$

Thus, approximately 159 units of work will be produced on the 30th day.

(b) When Q = 140, we have

$$140 = 200 - 120 e^{-0.036t}$$

$$0.5 = \frac{60}{120} = e^{-0.036t}$$

Taking the natural logarithm of both sides, we have

$$-0.036t = \ln 0.5$$

$$t = -\frac{1}{0.036} \ln 0.5 \approx 19.25$$

Thus, on the 20th day, 140 units of work will be produced.

15. We have the differential equation

$$\frac{dv}{dt} = \frac{k}{w} = (\frac{32w}{k} - v)$$

whose solution is

$$v = \frac{32w}{k}(1 - e^{-kt/w}) \text{ since } v_0 = 0$$

Since w = 220 and k = 33, we obtain

$$v = 32\frac{(220)}{33}[1 - e^{-(33)t/220}]$$

$$v = 32(\frac{20}{3})[1 - e^{-3t/20}]$$

(a) The velocity after t seconds is

$$v = \frac{640}{3}[1 - e^{-3t/20}]$$

(b) At t = 10, we have

$$v = \frac{640}{3}[1 - e^{-3(10)/20}] = 165.7 \text{ ft./sec.}$$

(c) The terminal velocity is

$$\frac{32w}{k} = \frac{32(220)}{33} = \frac{640}{3} = 213\frac{1}{3} \text{ ft./sec.}$$

19. (a) The logistic growth equation is

$$\frac{dP}{dt} = kP(L - P)$$

whose solution is

$$P = \frac{L}{1 + Ce^{-Lkt}}$$

The maximum population is L = 21,000. So

$$P = \frac{21000}{1 + Ce^{-21000kt}}$$

At $t = 0$, $P = 1000$. Thus

$$1000 = \frac{21000}{1 + Ce^{(-21000)(k)(0)}} = \frac{21000}{1 + C}$$

Hence

$$1 + C = \frac{21000}{1000} = 21$$

or

$$C = 20.$$

Then

$$P = \frac{21000}{1 + 20e^{-21000kt}}$$

At $t = 4$, $P = 3000$. Thus

$$3000 = \frac{21000}{1 + 20e^{-21000(4)k}}$$

$$1 + 20e^{-84000k} = \frac{21000}{3000} = 7$$

$$20e^{-84000k} = 6$$

$$e^{-84000k} = \frac{6}{20} = 0.3$$

Taking the natural logarithm of both sides, we have

$$-84000k = \ln 0.3$$

$$k = -\frac{1}{84000} \ln 0.3 = 0.0000143 = 0.14333 \times 10^{-4}$$

After t hours, the size of the population will be

$$P = \frac{21000}{1 + 20e^{-21000(0.143333 \times 10^{-4})t}} = \frac{21000}{1 + 20e^{-0.301t}}$$

At $t = 5$, we have

$$P = \frac{21000}{1 + 20e^{-0.301(5)}} = 3860$$

(b) At $P = 10{,}000$, we have

$$10{,}000 = \frac{21000}{1 + 20e^{-0.301t}}$$

$$1 + 20e^{-0.301t} = \frac{21000}{10000} = 2.1$$

$$20e^{-0.301t} = 1.1$$

$$e^{-0.301t} = \frac{1.1}{20}$$

Taking the natural logarithm of both sides, we have

$$-0.301t = \ln \frac{1.1}{20}$$

$$t = -\frac{1}{0.301} \ln \frac{1.1}{20} = 9.64$$

It takes 9.64 hours for the population to grow 10,000 bacteria.

Review Exercises, (Page 388)

7. $\int \dfrac{1}{\sqrt[3]{x^2}} dx = \int x^{-2/3} dx = \dfrac{x^{1/3}}{1/3} + C = 3\sqrt[3]{x} + C$

9. $\int (\sqrt[3]{x^4} - \dfrac{4}{x} + e^x) dx = \int (x^{4/3} - \dfrac{4}{x} + e^x) dx$

$$= \dfrac{x^{7/3}}{7/3} - 4 \ln |x| + e^x + C$$

$$= \dfrac{3}{7} x^{7/3} - 4 \ln |x| + e^x + C$$

13. Let $u = x^3 + 2$ so that $du = 3x^2 dx$ and $x^2 dx = \dfrac{1}{3} du$. Then

$$\int x^2 (x^3 + 2)^8 dx = \int \dfrac{1}{3} u^8 du = \dfrac{1}{3} \int u^8 du$$

$$= \dfrac{1}{3} \cdot \dfrac{u^9}{9} + C = \dfrac{(x^3 + 2)^9}{27} + C$$

17. Let $u = e^{2x} + 1$ so that $du = 2e^{2x} dx$ and $e^{2x} dx = \dfrac{1}{2} du$. Then

$$\int \dfrac{5e^{2x}}{e^{2x} + 1} dx = \int \dfrac{5}{u} \dfrac{1}{2} du = \dfrac{5}{2} \int \dfrac{1}{u} du$$

$$= \dfrac{5}{2} \ln |u| + C$$

$$= \dfrac{5}{2} \ln |e^{2x} + 1| + C$$

$$= \dfrac{5}{2} \ln (e^{2x} + 1) + C$$

since $e^{2x} + 1 > 0$ for all x.

21. Let

$$u(t) = t \text{ and } v'(t) = \sqrt{t - 3} = (t - 3)^{1/2}$$

so

$$u'(t) = 1 \text{ and } v(t) = \frac{2}{3}(t-3)^{3/2}$$

The method of substitution was used to determine v(t). Now

$$\int t\sqrt{t-3}\, dt = \int u(t)\, v'(t)\, dt$$

$$= u(t)\, v(t) - \int v(t)\, u'(t)\, dt$$

$$= t \cdot \frac{2}{3}(t-3)^{3/2} - \int \frac{2}{3}(t-3)^{3/2} \cdot 1\, dt$$

$$= \frac{2t}{3}(t-3)^{3/2} - \frac{2}{3}\int (t-3)^{3/2}\, dt$$

$$= \frac{2t}{3}(t-3)^{3/2} - \frac{2}{3} \cdot \frac{2}{5}(t-3)^{5/2} + C$$

$$= \frac{2t}{3}(t-3)^{3/2} - \frac{4}{15}(t-3)^{5/2} + C$$

31. We use Formula 11

$$\int \frac{u^2}{au+b} = \frac{1}{a^3}[\frac{1}{2}(au+b)^2 - 2b(au+b) + b^2 \ln|au+b|] + C$$

with u = t, a = -4, and b = 5.

$$\int \frac{t^2}{5-4t}\, dt = \int \frac{t^2}{-4t+5}\, dt$$

$$= \frac{1}{(-4)^3}[\frac{1}{2}(-4t+5)^2 - 2(5)(-4t+5) + (5)^2 \ln|-4t+5|] + C$$

$$= \frac{-1}{64}[\frac{1}{2}(5-4t)^2 - 10(5-4t) + 25 \ln|5-4t|] + C$$

33. If $y = -2x^2$, then $y = -4x$. Substituting these expressions into the given equation, we obtain

$$xy' - 2y = x(-4x) - 2(-2x^2) = -4x^2 + 4x^2 = 0$$

Thus $y = -2x^2$ is a solution of $xy' - 2y = 0$

39. We write the differential equation as

$$3x\frac{dy}{dx} = 3x^2 + 2$$

Separation of variables yields

$$dy = \frac{3x^2 + 2}{3x} dx = (x + \frac{2}{3x}) dx$$

Integrating both sides we obtain

$$\int dy = \int (x + \frac{2}{3x}) dx$$

or

$$y = \frac{1}{2}x^2 + \frac{2}{3}\ln |x| + C$$

41. We write the differential equation as

$$\frac{dy}{dx} = \frac{\sqrt{x}}{\sqrt{y}}$$

Separation of variables yields

$$\sqrt{y}\, dy = \sqrt{x}\, dx$$

Integrating both sides we obtain

$$\int y^{1/2}\, dy = \int x^{1/2}\, dx$$

or

$$\frac{2}{3}y^{3/2} = \frac{2}{3}x^{3/2} + C_1$$

$$y^{3/2} = x^{3/2} + C$$

where $C = 3\frac{C_1}{2}$ is an arbitrary constant. Now $y = 9$ when $x = 4$ so that

$$9^{3/2} = 4^{3/2} + C$$

or C = 19. Thus,

$$y^{3/2} = x^{3/2} + 19$$

Solving for y, we obtain

$$y = (x^{3/2} + 19)^{2/3}$$

45. The slope of the tangent line is given by the derivative. Thus

$$\frac{dy}{dx} = 6x^3 \sqrt{x^4 + 9}$$

Separating variables yields

$$dy = 6x^3 \sqrt{x^4 + 9}\, dx$$

We integrate both sides to obtain

$$\int dy = \int 6x^3 \sqrt{x^4 + 9}\, dx$$

Let $u = x^4 + 9$ so that $du = 4x^3\, dx$. Then $x^3\, dx = \frac{1}{4} du$ and

$$\int 6x^3 \sqrt{x^4 + 9}\, dx = \int 6u^{1/2}\left(\frac{1}{4}\right) du = \frac{3}{2}\int u^{1/2}\, du$$

$$= \frac{3}{2} \cdot \frac{2}{3} u^{3/2} + C$$

$$= u^{3/2} + C$$

$$= (x^4 + 9)^{3/2} + C$$

We have

$$y = (x^4 + 9)^{3/2} + C$$

Since the point (2, 250) lies on the curve we substitute $x = 2$ and $y = 250$ into the equation.

$$250 = (2^4 + 9)^{3/2} + C$$

$$250 = 125 + C$$

$$125 = C$$

An equation for the curve is

$$y = (x^4 + 9)^{3/2} + 125$$

47. We have

$$Q(t) = \int Q'(t)\, dt$$

$$= \int 10te^{-0.5t}\, dt$$

$$= 10\int te^{-0.5t}\, dt$$

The integrand is of the form in Formula 40.

$$\int xe^{ax}\, dx = \frac{e^{ax}}{a^2}(ax - 1) + C$$

We use $x = t$ and $a = -0.5$ to obtain

$$Q(t) = 10 \cdot \frac{e^{-0.5t}}{(-0.5)^2}(-0.5t - 1) + C$$

$$= 40e^{-0.5t}\left(-\frac{1}{2}t - 1\right) + C$$

Since $Q(0) = 0$, we can determine the constant of integration. Now

$$Q(0) = 40e^{-0.5(0)}\left(-\frac{1}{2}(0) - 1\right) + C = 0$$

or

$$-40 + C = 0$$
$$C = 40$$

Thus,

$$Q(t) = 40\left[1 - \left(\frac{1}{2}t + 1\right)e^{-0.5t}\right]$$

so that

$$Q(8) = 40\left[1 - \left(\frac{1}{2}\cdot 8 + 1\right)e^{-0.5(8)}\right]$$

$$= 40[1 - 5e^{-4}]$$

6: Antidifferentiation   163

$$= 36.337$$

The amount of particulate matter discharged during the first 8 years of operation is approximately 36.337 tons.

Chapter Test, (Page 390)

1. Let $u = 7x + 2$ so that $du = 7\, dx$. Then

    $$dx = \frac{1}{7} du$$

    and

    $$\int (7x + 2)^5 \, dx = \int u^5 \left(\frac{1}{7}\right) du = \frac{1}{7} \int u^5 \, du$$

    $$= \frac{1}{7} \frac{u^6}{6} + C$$

    $$= \frac{1}{42} (7x + 2)^6 + C$$

3. Let $u = x^2 + 5$ so that $du = 2x\, dx$. Then

    $$x \, dx = \frac{1}{2} du$$

    and

    $$\int \frac{x}{x^2 + 5} \, dx = \int \frac{1/2 \, du}{u} = \frac{1}{2} \int \frac{du}{u}$$

    $$= \frac{1}{2} \ln |u| + C$$

    $$= \frac{1}{2} \ln |x^2 + 5| + C$$

    $$= \frac{1}{2} \ln (x^2 + 5) + C$$

    since $x^2 + 5 > 0$.

5. Let $u(t) = t$ and $v'(t) = \sqrt{t + 2} = (t + 2)^{1/2}$ so that

$u'(t) = 1$ and $v(t) = \int v'(t) \, dt = \int (t+2)^{1/2} \, dt = \frac{2}{3}(t+2)^{3/2}$

Then

$$\int t\sqrt{t+2} \, dt = \int u(t) \, v'(t) \, dt$$

$$= u(t) \, v(t) - \int v(t) \, u'(t) \, dt$$

$$= t[\frac{2}{3}(t+2)^{3/2}] - \int \frac{2}{3}(t+2)^{3/2} (1) \, dt$$

$$= \frac{2}{3} t \, (t+2)^{3/2} - \frac{2}{3} \cdot \frac{2}{5}(t+2)^{5/2} + C$$

$$= \frac{2}{3} t \, (t+2)^{3/2} - \frac{4}{15}(t+2)^{5/2} + C$$

7. Let $u = x^2$ so that $du = 2x \, dx$. Then

$$x \, dx = \frac{1}{2} du$$

and

$$\int 3x e^{x^2} \, dx = 3\int e^{x^2} x \, dx = 3\int e^u \, (\frac{1}{2}) \, du$$

$$= \frac{3}{2} \int e^u \, du$$

$$= \frac{3}{2} e^u + C$$

$$= \frac{3}{2} e^{x^2} + C$$

9. We use Formula 12

$$\int \frac{u \, du}{(au+b)^2} = \frac{1}{a^2}[\frac{b}{au+b} + \ln|au+b|] + C$$

with $u = x$, $a = 5$, and $b = -4$ to obtain

$$\int \frac{3x}{(5x-4)^2} dx = 3 \int \frac{x\,dx}{(5x-4)^2} = \frac{3}{(5)^2}\left[\frac{-4}{5x-4} + \ln|5x-4|\right] + C$$

$$= \frac{3}{25}\left[\frac{-4}{5x-4} + \ln|5x-4|\right] + C$$

11. We use Formula 17

$$\int \frac{du}{(au+b)(cu+d)} = \frac{1}{bc-ad} \ln\left|\frac{cu+d}{au+b}\right| + C$$

with $u = x, a = 1, b = -1, c = 1, d = 3$ to obtain

$$\int \frac{1}{x^2 + 2x - 3} dx = \int \frac{dx}{(x-1)(x+3)} = \frac{1}{(-1)(1) - (1)(3)} \ln\left|\frac{x+3}{x-1}\right| + C$$

$$= -\frac{1}{4} \ln\left|\frac{x+3}{x-1}\right| + C$$

13. We write the differential equation as

$$\frac{dy}{dx} = 4xy^2$$

Separation of variables yields

$$\frac{dy}{y^2} = 4x\,dx$$

Integrating both sides we obtain

$$\int \frac{dy}{y^2} = \int 4x\,dx$$

or

$$-\frac{1}{y} = 2x^2 + C$$

Substituting $y = 1$ when $x = 0$ gives

$$-1 = 0 + C \quad \text{or} \quad C = -1$$

Thus

$$-\frac{1}{y} = 2x^2 - 1$$

or

$$y = \frac{-1}{2x^2 - 1}$$

solves the initial value problem.

7 The Definite Integral

Key Ideas for Review

* The First Fundamental Theorem of Calculus: Suppose that f is a function that is continuous and nonnegative on the interval [a,b]. Let A(x) denote the area of the region under the curve y = f(x) over the interval [a,x]. Then A'(x) = f(x) for each x in [a,b].

* If f(x) is a continuous function on the interval [a,b]. we define the definite integral of f from a to b by

  $\int_a^b f(x)\, dx = F(b) - F(a)$

  where F(x) is any antiderivative of f(x).

* If f(x) is continuous and nonnegative on [a,b], then $\int_a^b f(x)\, dx$ represents the area of the region under the curve y = f(x) over the interval [a,b].

* Properties of the definite integral:

  $\int_a^a f(x)\, dx = 0$

  $\int_a^b kf(x)\, dx = k\int_a^b f(x)\, dx$

  $\int_a^b f(x)\, dx = -\int_b^a f(x)\, dx$

  $\int_a^b [f(x) \pm g(x)]\, dx = \int_a^b f(x)\, dx \pm \int_a^b g(x)\, dx$

  $\int_a^b f(x)\, dx = \int_a^c f(x)\, dx + \int_c^b f(x)\, dx$

* If f is a continuous function on [a,b], then the definite integral $\int_a^b f(x)\, dx$ represents the area of the region lying above [a,b] and below y = f(x) minus the area of the region lying below [a,b] and above y = f(x).

* If f and g are continuous functions with $f(x) \geq g(x)$ on [a,b], then the area of the region between the curves $y = f(x)$ and $y = g(x)$ over the interval [a,b] is $\int_a^b [f(x) - g(x)]\, dx$.

* If the rate of change $f'(x)$ of $f(x)$ with respect to x is continuous on [a,b], then the definite integral $\int_a^b f'(x)\, dx$ gives the change in the value of $f(x)$ as x varies from a to b.

* Divide the interval [a,b] into n subintervals of equal width $\Delta x = (b - a)/n$. Let x_k be an arbitrary point in the kth subinterval. The expression

$$f(x_1)\Delta x + f(x_2)\Delta x + \ldots + f(x_n)\Delta x$$

 is called a **Riemann sum**.

* If $f(x)$ is continuous and nonnegative on [a,b], then the limit of the Riemann sums as $n \longrightarrow +\infty$ is $\int_a^b f(x)\, dx$.

* The average value of $f(x)$ over [a,b] is given by

$$y_{ave.} = \frac{1}{b-a} \int_a^b f(x)\, dx.$$

* Let f be a continuous function for $x \geq a$. The improper integral of f from a to $+\infty$ denoted by $\int_a^{+\infty} f(x)\, dx$, is defined by

$$\int_a^{+\infty} f(x)\, dx = \lim_{b \to +\infty} \int_a^b f(x)\, dx$$

 if this limit exists. An improper integral is said to converge or diverge.

* Similarly, we define the improper integrals

$$\int_{-\infty}^b f(x)\, dx = \lim_{a \to -\infty} \int_a^b f(x)\, dx$$

 and

$$\int_{-\infty}^{+\infty} f(x)\, dx = \int_{-\infty}^0 f(x)\, dx + \int_0^{+\infty} f(x)\, dx$$

7: *The Definite Integral* 169

The following items all pertain to the optional sections of Chapter 7.

* If a continuous stream of income generates income at the rate of f(t) dollars per year t years from now, then

    -- the total amount of the flow from time t_1 to time t_2 is

    $$\text{total flow} = \int_{t_1}^{t_2} f(t)\, dt$$

    -- the total future value of the income stream after T years, inverted at the continuous compound interest rate r, is

    $$S(T) = \int_0^T f(t) e^{r(T-t)}\, dt$$
    $$= e^{rT} \int_0^T f(t) e^{-rt}\, dt.$$

    -- the present value of the first T years of income is

    $$P(T) = \int_0^T f(t) e^{-rt}\, dt.$$

    assuming that the prevailing continuous compound interest rate is r.

* Consumers' and Producers' Surplus. If p = D(x) is the demand function and p = S(x) is the supply function for a certain commodity, then when $\tilde{x}$ units are supplied (and sold) on the market and when the selling price is $\tilde{p}$.

    $$CS = \int_0^{\tilde{x}} [D(x) - \tilde{p}]\, dx,$$

and

$$PS = \int_0^{\tilde{x}} [\tilde{p} - S(x)]\, dx.$$

* The present value of an endless stream of income:

    $$P(+\infty) = \int_0^{+\infty} f(t) e^{-rt}\, dt.$$

* Rectangular approximations of $\int_a^b f(x)\, dx$:

    $$L_n = \Delta x [y_0 + y_1 + y_2 + \ldots + y_{n-1}]$$
    $$R_n = \Delta x [y_1 + y_2 + y_3 + \ldots + y_n]$$

$$M_n = \Delta x[\tilde{y}_1 + \tilde{y}_2 + \tilde{y}_3 + \ldots + \tilde{y}_n]$$

where $\Delta x = \dfrac{b-a}{n}$, $y_i = f(x_i)$ and

$$\tilde{y}_i = f(\dfrac{x_{i-1} + x_i}{2})$$

* Trapezoidal approximation of $\int_a^b f(x)\,dx$:

$$T_n = \dfrac{\Delta x}{2}[y_0 + 2y_1 + 2y_2 + \ldots + 2y_{n-1} + y_n]$$

where $\Delta x = \dfrac{b-a}{n}$ and $y_i = f(x_i)$.

Exercise Set 7.1, (Page 398)

3. (a) We graph $y = 2x$.

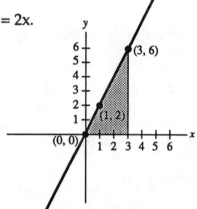

(b) We seek the area of the triangular region under the graph. The area of a triangle is one-half the base times the height.

$$A = \dfrac{1}{2}(\text{base})(\text{height}) = \dfrac{1}{2}(x)(y)$$

Since the interval is [0,3], the base is $x = 3$, and the height is $y = 2x = 2(3) = 6$. Thus the area is

$$A = \dfrac{1}{2}(3)(6) = 9$$

(c) The desired area is

$$A(3) = \int_0^3 f(x)\, dx = \int_0^3 2x\, dx$$

$$= 2\int_0^3 x\, dx = \frac{2x^2}{2}\Big]_0^3 = x^2$$

$$= (3)^2 - 0 = 9$$

9.

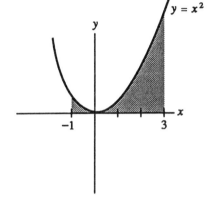

The desired area is

$$A(3) = \int_{-1}^3 x^2\, dx = \frac{x^3}{3}\Big]_{-1}^3 = \frac{3^3}{3} - \frac{(-1)^3}{3} = 9 + \frac{1}{3} = \frac{28}{3}$$

15.

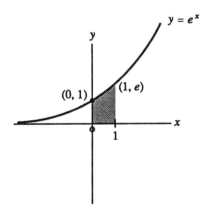

The desired area is

$$A(1) = \int_0^1 f(x)\, dx = \int_0^1 e^x\, dx = e^x\Big]_0^1 = e^1 - e^0 = e - 1$$

17.

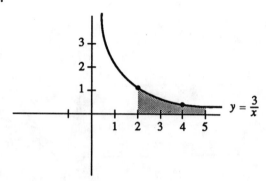

The desired area is

$$A(5) = \int_2^5 \frac{3}{x}\, dx = 3 \ln |x|\Big]_2^5$$

$$= 3(\ln 5 - \ln 2) = 3 \ln \frac{5}{2}$$

23. $\int_0^2 e^t\, dt = e^t\Big]_0^2 = e^2 - e^0 = e^2 - 1$

27. $\int_1^4 \sqrt{x}\, dx = \int_1^4 x^{1/2}\, dx = \frac{2}{3} x^{3/2}\Big]_1^4$

$$= \frac{2}{3}(4)^{3/2} - \frac{2}{3}(1)^{3/2}$$

$$= \frac{2}{3}(8) - \frac{2}{3}(1) = \frac{2}{3}(8-1) = \frac{14}{3}$$

31. The desired area is

$$A(2) = \int_{-1}^2 e^{-x}\, dx = -1e^{-x}\Big]_{-1}^2 = -1(e^{-2} - e^{-(-1)})$$

$$= -1[\frac{1}{e^2} - e] = e - \frac{1}{e^2}$$

Exercise Set 7.2, (Page 412)

11. $\int_{-1}^{2}(3x^3 - 2x^2 + x)\,dx = \left(\dfrac{3x^4}{4} - \dfrac{2x^3}{3} + \dfrac{x^2}{2}\right)\Big]_{-1}^{2}$

$= (\dfrac{3}{4}(2)^4 - \dfrac{2}{3}(2)^3 + \dfrac{(2)^2}{2}) - (\dfrac{3}{4}(-1)^4 - \dfrac{2}{3}(-1)^3 + \dfrac{(-1)^2}{2})$

$= \dfrac{3}{4}(16 - 1) - \dfrac{2}{3}(8 + 1) + \dfrac{1}{2}(4 - 1)$

$= \dfrac{45}{4} - 6 + \dfrac{3}{2} = \dfrac{45 - 24 + 6}{4} = \dfrac{27}{4}$

17. $\int_{0}^{2}(e^{-2t} - 3t^2)\,dt = \left(-\dfrac{1}{2}e^{-2t} - t^3\right)\Big]_{0}^{2}$

$= \left(\dfrac{1}{2}e^{-2t} - t^3\right)\Big]_{2}^{0}$

$= (\dfrac{1}{2}e^{-2(0)} + (0)^3) - (\dfrac{1}{2}e^{-2(2)} + (2)^3)$

$= \dfrac{1}{2} - (\dfrac{1}{2}e^{-4} + 8)$

$= -\dfrac{1}{2}e^{-4} - \dfrac{15}{2} = -\dfrac{1}{2}(e^{-4} + 15)$

19. $\int_{1}^{8}(u^{-1/3} + 2u^3)\,du = \left(\dfrac{3}{2}u^{2/3} + \dfrac{u^4}{2}\right)\Big]_{1}^{8}$

$= (\dfrac{3}{2}(8)^{2/3} + \dfrac{(8)^4}{2}) - (\dfrac{3}{2}(1)^{2/3} + \dfrac{(1)^4}{2})$

$= (\dfrac{3}{2}(4) + \dfrac{4096}{2}) - (\dfrac{3}{2} + \dfrac{1}{2})$

$= 6 + 2048 - 2 = 2052$

23. $\int_{1}^{4}(\dfrac{1}{2\sqrt{x}} + 5\sqrt{x})\,dx = \int_{1}^{4}(\dfrac{1}{2}x^{-1/2} + 5x^{1/2})\,dx$

$= \left(x^{1/2} + \dfrac{5x^{3/2}}{3/2}\right)\Big]_{1}^{4}$

$$= \left(x^{1/2} + \frac{10}{3}x^{3/2}\right)\bigg]_1^4$$

$$= \left((4)^{1/2} + \frac{10}{3}(4)^{3/2}\right) - \left((1)^{1/2} + \frac{10}{3}(1)^{3/2}\right)$$

$$= (2 + \frac{10}{3}(8)) - (1 + \frac{10}{3})$$

$$= 2 + \frac{80}{3} - 1 - \frac{10}{3} = 1 + \frac{70}{3} = \frac{73}{3}$$

31. The desired area A is the sum of the area under the curve $y = 1 + x$ over the interval [-1, 0] and the area under the curve $y = x^2 + 1$ over the interval [0,2]. Hence

$$A = \int_{-1}^{0} (1 + x)\,dx + \int_{0}^{2} (x^2 + 1)\,dx$$

$$= \left(x + \frac{1}{2}x^2\right)\bigg]_{-1}^{0} + \left(\frac{x^3}{3} + x\right)\bigg]_0^2$$

$$= (0 + \frac{1}{2}(0)^2) - (-1 + \frac{1}{2}(-1)^2) + (\frac{(2)^3}{3} + 2) - (\frac{(0)^3}{3} + 0)$$

$$= 0 - (-1 + \frac{1}{2}) + (\frac{8}{3} + 2) - 0$$

$$= \frac{1}{2} + \frac{8}{3} + 2 = \frac{3 + 16 + 12}{6} = \frac{31}{6}$$

35. We first sketch the region in question

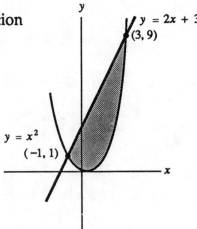

The x-coordinates of the points of intersection are found by solving

$x^2 = 2x + 3$
$x^2 - 2x - 3 = 0$
$(x + 1)(x - 3) = 0$
$x + 1 = 0 \quad x - 3 = 0$
$\quad x = -1 \quad\quad x = 3$

Thus we shall integrate over the interval [-1,3]. The area is

$$A = \int_{-1}^{3}[(2x+3)-x^2]\,dx = \int_{-1}^{3}(2x+3-x^2)\,dx = x^2 + 3x - \frac{x^3}{3}\Big]_{-1}^{3}$$

$$= [(3)^2 + 3(3) - \frac{(3)^3}{3}] - [(-1)^2 + 3(-1) - \frac{(-1)^3}{3}]$$

$$= (9+9-9) - (1-3+\frac{1}{3}) = \frac{32}{3}$$

37. We first sketch the region in question

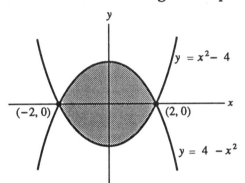

The x-coordinates of the points of intersection are found by solving

$x^2 - 4 = 4 - x^2$
$2x^2 = 8$
$x^2 = 4$
$x = \pm 2$

Thus we shall integrate over the interval [-2, 2]. The area is

$$A = \int_{-2}^{2}[(4-x^2)-(x^2-4)]\,dx = \int_{-2}^{2}(4-x^2-x^2+4)\,dx$$

$$= \int_{-2}^{2}(8-2x^2)\,dx = 8x - \frac{2}{3}x^3\Big]_{-2}^{2}$$

$$= [8(2) - \frac{2}{3}(2)^3] - [8(-2) - \frac{2}{3}(-2)^3]$$

$$= (16 - \frac{16}{3}) - (-16 + \frac{16}{3}) = \frac{64}{3}$$

45. We first sketch the region in question.

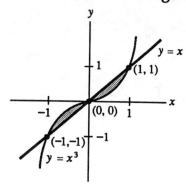

The x-coordinates of the points of intersection are found by solving

$$x^3 = x$$
$$x^3 - x = 0$$
$$x(x^2 - 1) = 0$$
$$x = 0 \qquad x = \pm 1$$

We shall integrate separately over the intervals [-1,0] and [0,1]. We note that

upper boundary curve: $\quad y = \begin{cases} x^3 & -1 \leq x \leq 0 \\ x & 0 \leq x \leq 1 \end{cases}$

lower boundary curve: $\quad y = \begin{cases} x & -1 \leq x \leq 0 \\ x^3 & 0 \leq x \leq 1 \end{cases}$

Thus

$$A = \int_{-1}^{0} (x^3 - x)\,dx + \int_{0}^{1} (x - x^3)\,dx$$

$$= \frac{x^4}{4} - \frac{x^2}{2} \Big]_{-1}^{0} + = \frac{x^2}{2} - \frac{x^4}{4} \Big]_{0}^{1}$$

7: The Definite Integral 177

$$= [0 - \frac{(-1)^4}{4} - \frac{(-1)^2}{2}] + [\frac{(1)^2}{2} - \frac{(1)^4}{4} - 0]$$

$$= -(\frac{1}{4} - \frac{1}{2}) + (\frac{1}{2} - \frac{1}{4}) = \frac{1}{2}$$

Exercise Set 7.3, (Page 423)

5. (a) The amount of oil produced in the first four years of operation is given by

$$\int_0^4 R'(t) \, dt$$

since $R(0) = 0$. Now

$$\int_0^4 R'(t) \, dt = \int_0^4 (40 + \frac{3}{2}t - \frac{1}{2}t^3) \, dt$$

$$= (40t + \frac{3}{4}t^2 = \frac{1}{8}t^4) \Big]_0^4$$

$$= (40(4) + \frac{3}{4}(4)^2 - \frac{1}{8}(4)^4) - 0$$

$$= 160 + 12 - 32 = 140$$

Thus 140 million barrels are produced in the first four years.

(b) The amount of oil produced in the combined third and fourth years of operation is given by

$$\int_2^4 R'(t) \, dt = \int_2^4 (40 + \frac{3}{2}t - \frac{1}{2}t^3) \, dt$$

$$= 40t + \frac{3}{4}t^2 = \frac{1}{8}t^4 \Big]_2^4$$

$$= [40(4) + \frac{3}{4}(4)^2 - \frac{1}{8}(4)^4] - [40(2) + \frac{3}{4}(2)^2 - \frac{1}{8}(2)^4]$$

$$= (160 + 12 - 32) - (80 + 3 - 2) = 140 - 81 = 59$$

Thus, 59 million barrels are produced in the combined third and fourth years.

9. (a) We need to compute the change in revenue from x = 20 to x = 40. Thus, this change is given by

$$\int_{20}^{40} R'(x)\, dx = \int_{20}^{40} (400 + 0.1x)\, dx$$

$$= \left(400 + 0.1\frac{x^2}{2}\right)\Big]_{20}^{40}$$

$$= (400(40) + 0.05(40)^2) - (400(20) + 0.05(20)^2)$$

$$= 16{,}080 - 8020 = 8060$$

The revenue increases by $8060 as x changes from 20 to 40.

(b) The revenue received from the manufacture and sale of 50 pairs of skis is

$$\int_0^{50} R'(x)\, dx = \int_0^{50} (400 + 0.1x)\, dx$$

$$= \left(400 + 0.05x^2\right)\Big]_0^{50}$$

$$= (400)(50) + 0.05(50)^2) - (400(0) + 0.05(0)^2)$$

$$= 20{,}125$$

The revenue from the sale of 50 pairs of skis is $20,125.

(c) The change in cost from x = 0 to x = 50 is

$$\int_0^{50} C'(x)\, dx = \int_0^{50} (400 - 0.2x)\, dx$$

$$= \left(400 + 0.1x^2\right)\Big]_0^{50}$$

$$= (400(50) - 0.1(50)^2) - (400(0) - 0.1(0)^2)$$

$$= 19{,}750$$

The cost of producing 50 pairs of skis is the fixed cost plus the change in cost from $x = 0$ to $x = 50$; that is,

$$\$200 + \$19{,}750 = \$19{,}950$$

(d) The profit received from the manufacture and sale of 50 pairs of skis is the difference between the associated revenue and the cost. Thus, this profit is

$$\$20{,}125 - \$19{,}950 = \$175$$

15. (a) The change in velocity is given by

$$v(t_2) - v(t_1) = \int_{t_1}^{t_2} a(t)\, dt$$

During the first three seconds, we have

$$v(3) - v(0) = \int_0^3 32\, dt = 32t \Big]_0^3$$

$$= 32(3) - 32(0) = 96 \text{ ft./sec.}$$

During the next three seconds, we have

$$v(3) - v(0) = \int_3^6 32\, dt = 32t \Big]_3^6$$

$$= 32(6) - 32(3) = 96 \text{ ft./sec.}$$

(b) Since the stone is dropped, $v(0) = 0$. Thus

$$v(3) - v(0) = \int_0^t 32\, dt = 32t \Big]_0^t$$

The change in the distance is given by

$$x(t_2) - x(t_1) = \int_{t_1}^{t_2} v(t)\, dt$$

During the first three seconds, we have

$$x(3) - x(0) = \int_0^3 32t\, dt = \frac{32t^2}{2} \Big]_0^3$$

$$= 16(3)^2 - 16(0)^2 = 144 \text{ ft.}$$

During the next three seconds, we have

$$x(6) - x(3) = \int_3^6 32t \, dt = \frac{32t^2}{2}\Big]_3^6$$

$$= 16(6)^2 - 16(3)^2 = 576 - 144 = 432 \text{ ft.}$$

17. The rate of change of the revenue is given by

$$R'(t) = 95 - \frac{1}{4}t - \frac{3}{2}t^2$$

(a) The total earnings over the first four years will be

$$R(4) - R(0) = \int_0^4 (95 - \frac{1}{4}t - \frac{3}{2}t^2) \, dt$$

$$= 95t - \frac{1}{8}t^2 - \frac{1}{2}t^3 \Big]_0^4$$

$$= [95(4) - \frac{1}{8}(4)^2 - \frac{1}{2}(4)^3] - [95(0) - \frac{1}{8}(0)^2 - \frac{1}{2}(0)^3]$$

$$= 380 - 2 - 32 - 0 = 346 \text{ thousands of dollars}$$

(b) The total earnings over the next four years will be

$$R(8) - R(4) = \int_4^8 (95 - \frac{1}{4}t - \frac{3}{2}t^2) \, dt$$

$$= 95t - \frac{1}{8}t^2 - \frac{1}{2}t^3$$

$$= [95(8) - \frac{1}{8}(8)^2 - \frac{1}{2}(8)^3] - [95(4) - \frac{1}{8}(4)^2 - \frac{1}{2}(4)^3]$$

$$= (760 - 8 - 256) - (380 - 2 - 32) = 496 - 346$$

$$= 150 \text{ thousands of dollars}$$

Exercise Set 7.4, (Page 437)

3. We divide the interval [0,3] into n subintervals, each of width

$$\Delta x = \frac{3-0}{n} = \frac{3}{n}$$

We denote by x_k the right-hand endpoint of the kth subinterval for $k = 1, 2, \ldots, n$. Thus,

$$x_1 = 1\left(\frac{3}{n}\right) = \frac{3}{n},$$
$$x_2 = 2\left(\frac{3}{n}\right) = \frac{6}{n},$$
$$x_3 = 3\left(\frac{3}{n}\right) = \frac{9}{n},$$
$$\vdots$$
$$x_n = n\left(\frac{3}{n}\right) = 3$$

Then, since $f(x) = x + 2$,

the height of the first rectangle $= f(x_1) = f\left(\frac{3}{n}\right) = \frac{3}{n} + 2$

the height of the second rectangle $= f(x_2) = f\left(\frac{6}{n}\right) = \frac{6}{n} + 2$

the height of the third rectangle $= f(x_3) = f\left(\frac{9}{n}\right) = \frac{9}{n} + 2$.

$$\vdots$$

the height of the nth rectangle $= f(x_n) = f\left(\frac{3n}{n}\right) = \frac{3n}{n} + 2$

The Riemann sum consisting of the sum of the areas of the approximating rectangles is

$$R = \left(\frac{3}{n} + 2\right)\left(\frac{3}{n}\right) + \left(\frac{6}{n} + 2\right)\left(\frac{3}{n}\right) + \left(\frac{9}{n} + 2\right)\left(\frac{3}{n}\right) + \ldots + \left(\frac{3n}{n} + 2\right)\left(\frac{3}{n}\right)$$

$$= \frac{3}{n}(\frac{3}{n} + \frac{6}{n} + \frac{9}{n} + \ldots + \frac{3n}{n}) + \frac{3}{n}(2 + 2 + \ldots + 2)$$

$$= (\frac{3}{n})(\frac{3}{n})(1 + 2 + 3 + \ldots + n) + \frac{3}{n}(2 + 2 + \ldots + 2)$$

Now

$$1 + 2 + 3 + \ldots + n = \frac{n(n+1)}{2}$$

and since there are n identical terms in the sum $2 + 2 + \ldots + 2$, we obtain

$$2 + 2 + \ldots + 2 = 2n$$

Thus, the Riemann sum reduces to

$$R = \frac{9}{n^2}\frac{n(n+1)}{2} + (\frac{3}{n})2n$$

$$= \frac{9}{2}\frac{n+1}{n} + 6 = \frac{9}{2}(1 + \frac{1}{n}) + 6$$

As the number of subintervals increases, $n \to +\infty$. Then

$$\lim_{n \to +\infty} [\frac{9}{2}(1 + \frac{1}{n}) + 6] = \frac{9}{2}(1 + 0) + 6 = \frac{9}{2} + 6 = \frac{21}{2}$$

We conclude that the area under the curve $y = x + 2$ over the interval [0,3] is 21/2.

11. We divide the interval [0,1] into n subintervals, each of width

$$\Delta x = \frac{1 - 0}{n} = \frac{1}{n}$$

We denote by x_k the right-hand endpoint of the kth subinterval for $k = 1, 2, \ldots, n$. Thus,

$$x_1 = 1(\frac{1}{n}) = \frac{1}{n},$$

$$x_2 = 2(\frac{1}{n}) = \frac{2}{n},$$
$$x_3 = 3(\frac{1}{n}) = \frac{3}{n},$$
$$\vdots$$
$$x_n = n(\frac{1}{n}) = \frac{n}{n}$$

Then, since $f(x) = 5x$,

the height of the first rectangle $= f(x_1) = f(\frac{1}{n}) = 5(\frac{1}{n})$

the height of the second rectangle $= f(x_2) = f(\frac{2}{n}) = 5(\frac{2}{n})$

the height of the third rectangle $= f(x_3) = f(\frac{3}{n}) = 5(\frac{3}{n})$

$$\vdots$$

the height of the nth rectangle $= f(x_n) = f(\frac{n}{n}) = 5(\frac{n}{n})$

The Riemann sum consisting of the sum of the areas of the approximating rectangles is

$$R = 5(\frac{1}{n}) \cdot \frac{1}{n} + 5(\frac{2}{n}) \cdot \frac{1}{n} + 5(\frac{3}{n}) \cdot \frac{1}{n} + 5(\frac{n}{n}) \cdot \frac{1}{n}$$

$$= \frac{5}{n^2}(1 + 2 + 3 + \ldots + n)$$

$$= \frac{5}{n^2}[\frac{n(n+1)}{2}]$$

$$= \frac{5}{2}(1 + \frac{1}{n})$$

As the number of subintervals increases, $n \longrightarrow +\infty$. Then

$$\lim_{n \to +\infty} \frac{5}{2}(1 + \frac{1}{n}) = \frac{5}{2}(1 + 0) = \frac{5}{2}.$$

so that

$$\int_0^1 5x \, dx = \frac{5}{2}$$

15. We divide the interval [0,2] into n subintervals, each of width

$$\Delta x = \frac{2-0}{n} = \frac{2}{n}$$

We denote by x_k the right-hand endpoint of the kth subinterval for $k = 1, 2, \ldots, n$. Thus,

$$x_1 = 1(\frac{2}{n}) = \frac{2}{n},$$
$$x_2 = 2(\frac{2}{n}) = \frac{4}{n},$$
$$x_3 = 3(\frac{2}{n}) = \frac{6}{n},$$
$$\vdots$$
$$x_n = n(\frac{2}{n}) = \frac{2n}{n}$$

Then, since $f(x) = 2x^2 + 1$,

the height of the first rectangle $= f(x_1) = f(\frac{2}{n}) = 2(\frac{2}{n})^2 + 1$

the height of the second rectangle $= f(x_2) = f(\frac{4}{n}) = 2(\frac{4}{n})^2 + 1$

the height of the third rectangle $= f(x_3) = f(\frac{6}{n}) = 2(\frac{6}{n})^2 + 1$

$$\vdots$$

the height of the nth rectangle = $f(x_n) = f(\frac{2n}{n}) = 2(\frac{2n}{n})^2 + 1$

The Riemann sum consisting of the sum of the areas of the approximating rectangles is

$$R = [2(\frac{2}{n})^2 + 1]\frac{2}{n} + [2(\frac{2}{n})^2(2)^2 + 1]\frac{2}{n} + [2(\frac{2}{n})^2(3)^2 + 1]\frac{2}{n} + \ldots$$
$$[(2(\frac{2}{n})^2 n^2 + 1]\frac{2}{n}$$

$$= \frac{2}{n}[2(\frac{2}{n})^2(1 + 2^2 + 3^2 + \ldots + n^2) + (1 + 1 + 1 + \ldots + 1)]$$

Now

$$1 + 2^2 + 3^2 + \ldots + n^2 = \frac{n(n+1)(2n+1)}{6}$$

and since there are n identical terms in the sum $1 + 1 + \ldots + 1$, we obtain

$$1 + 1 + \ldots + 1 = n$$

Thus, the Reimann sum reduces to

$$R = \frac{2}{n}[\frac{8}{n^2}\frac{n(n+1)(2n+1)}{6} + n]$$

$$= \frac{8}{3n^2}(n+1)(2n+1) + \frac{2}{n} \cdot n$$

$$= \frac{8}{3}(1 + \frac{1}{n})(2 + \frac{1}{n}) + 2$$

As the number of subintervals increases, $n \longrightarrow +\infty$. Then,

$$\lim_{n \to +\infty} \frac{8}{3}[(1 + \frac{1}{n})(2 + \frac{1}{n}) + 2] = \frac{8}{3}(1 + 0)(2 + 0) + 2 = \frac{16}{3} + 2 = \frac{22}{3}$$

Thus

$$\int_0^2 (2x^2 + 1)\,dx = \frac{22}{3}$$

17. Here $f(x) = \sqrt{x}$ and $[a,b] = [0,3]$, so that

$$y_{ave.} = \frac{1}{b-a} \int_a^b f(x)\, dx = \frac{1}{3-0} \int_0^3 \sqrt{x}\, dx$$

$$= \frac{1}{3} \cdot \frac{2}{3} x^{3/2} \Big]_0^3$$

$$= \frac{2}{9} \cdot 3^{3/2} - \frac{2}{9} \cdot 0^{3/2}$$

$$= \frac{2}{9} \cdot 3^{3/2} = \frac{2}{3^2} \cdot 3^{3/2} = 2 \cdot 3^{3/2-2} = 2 \cdot 3^{-1/2} = \frac{2}{\sqrt{3}}$$

21. The average temperature between 3 P.M. (t = 3) and 8 P.M. (t = 8) is

$$\frac{1}{8-3} \int_3^8 h(t)\, dt = \frac{1}{5} \int_3^8 \left(\frac{5}{36} t^2 - \frac{10}{3} t + 20 \right) dt$$

$$= \int_3^8 \left(\frac{1}{36} t^2 - \frac{2}{3} t + 4 \right) dt$$

$$= \left(\frac{1}{36} \cdot \frac{t^3}{3} - \frac{2}{3} \cdot \frac{t^2}{2} + 4t \right) \Big]_3^8$$

$$= \frac{1}{108} (8^3 - 3^3) - \frac{1}{3}(8^2 - 3^2) + 4(8 - 3)$$

$$= \frac{1}{108} (512 - 27) - \frac{1}{3}(64 - 9) + 4(5)$$

$$= \frac{485}{108} - \frac{55}{3} + 20 = \frac{665}{108} = 6.16°$$

Thus, the average temperature is about 6.16° Fahrenheit.

Exercise Set 7.5, (Page 449)

3. The total future value of an income stream producing income t years from time t = 0 at the rate of f(t) dollars per year, invested at the prevailing rate of return r, T years from time t = 0 is

$$S(T) = \int_0^T f(t)\, e^{r(T-t)}\, dt$$

The revenues are received after t years at the rate of f(t) = 1000 - 50t dollars per year, and the prevailing rate of return is r = 0.10. We want to find the future value S(9). We have

$$S(9) = \int_0^9 (1000 - 50t)e^{0.10(9-t)} dt$$

$$= e^{(0.10)(9)} [1000 \int_0^9 e^{-0.10t} dt - 50 \int_0^9 te^{-0.10t} dt]$$

The first integral is evaluated using (6) and the second integral using Equation (40) in the Table of Integrals.

$$\int te^{-0.10t} dt = \frac{e^{-0.10t}}{(-0.10)^2} (-0.10t - 1)$$

$$= -\frac{e^{-0.10t}}{(-0.10)^2} (0.10t + 1)$$

Thus

$$S(9) = e^{(0.10)(9)}[1000(\frac{e^{-0.10t}}{(-0.10)} + 50\frac{e^{-0.10t}}{(0.10)^2}(0.10t+1)]\Big|_0^9$$

$$= e^{0.90}[e^{-0.10t}(-10,000 + \frac{50t}{(0.10)} + \frac{50}{(0.10)^2})]\Big|_0^9$$

$$= e^{0.90} [e^{-0.90} (-10,000 + 4500 + 5000) - e^0 (-10,000 + 5000)]$$

$$= e^0 (-500) - e^{0.90} (-5000)$$

$$= -500 + 12298$$

$$= \$11,798$$

7. If $p = D(x) = 30 - \frac{x^2}{2}$, then when $x = \tilde{x} = 6$, we have $p = \tilde{p} = 30 - (6)^2/2 = 30 - 18 = 12$. Thus the consumers' surplus is

$$CS = \int_0^{\tilde{x}} (D(x) - \tilde{p}) dx$$

$$CS = \int_0^6 (30 - \frac{x^2}{2} - 12) dx = \int_0^6 (18 - \frac{x^2}{2}) dx$$

$$= 18x - \frac{x^3}{6}\Big]_0^6 = [(18)(6) - \frac{(6)^3}{6}] - [(18)(0) - \frac{(0)^3}{6}]$$

$$= 108 - 36 - 0 = 72$$

9. If $p = S(x) = 3 + \frac{x^2}{8}$, then when $p = \tilde{p} = 21$, we have

$$21 = 3 + \frac{x^2}{8}$$

$$18 = \frac{x^2}{8}$$

$$144 = x^2$$

$$\pm 12 = x$$

Thus $x = \tilde{x} = 12$. The producers' surplus is

$$PS = \int_0^{\tilde{x}} [\tilde{p} - S(x)]\, dx$$

$$PS = \int_0^{12} [21 - (3 + \frac{x^2}{8})]\, dx = \int_0^{12} (18 - \frac{x^2}{8})\, dx$$

$$= 18x - \frac{x^3}{24}\Big]_0^{12} = [(18)(12) - \frac{(12)^3}{24}] - [(18)(0) - \frac{(0)^3}{24}]$$

$$= 216 - 72 - 0 = 144$$

13. We first find the equilibrium price by equating $S(x)$ and $D(x)$.

$$x^2 + 1 = 9 - 2x$$
$$x^2 + 2x - 8 = 0$$
$$(x - 2)(x + 4) = 0$$
$$x = 2 \quad x = -4$$

Since the number of units cannot be negative, we have $\tilde{x} = x_E = 2$. Now, using either the supply or the demand function, we obtain $P_E = 5$. Thus

$$CS = \int_0^2 [(9 - 2x) - 5]\, dx = \int_0^2 (4 - 2x)\, dx = 4x - x^2\Big]_0^2$$

$$= [4(2) - (2)^2] - [4(0) - (0)^2] = 8 - 4 = 4$$

and

$$PS = \int_0^2 [5 - (x^2 + 1)] \, dx = \int_0^2 (4 - x^2) \, dx = 4x - \frac{x^3}{3} \Big]_0^2$$

$$= [(4)(2) - \frac{(2)^3}{3}] - [4(0) - \frac{(0)^3}{3}] = 8 - \frac{8}{3} = \frac{16}{3}$$

Exercise Set 7.6, (Page 458)

1. $\int_0^{+\infty} \frac{1}{x^2} \, dx = \lim_{b \to +\infty} \int_2^b \frac{1}{x^2} \, dx$

Now

$$\int_2^b \frac{1}{x^2} \, dx = \int_2^b x^{-2} \, dx = \frac{x^{-1}}{-1} \Big]_2^b = -\frac{1}{x} \Big]_2^b$$

$$= (-\frac{1}{b}) - (-\frac{1}{2}) = \frac{1}{2} - \frac{1}{b}$$

Then

$$\int_2^{+\infty} \frac{1}{x^2} \, dx = \lim_{b \to +\infty} (\frac{1}{2} - \frac{1}{b}) = \frac{1}{2} - 0 = \frac{1}{2}$$

3. $\int_0^{+\infty} e^{-2x} \, dx = \lim_{b \to +\infty} \int_0^b e^{-2x} \, dx$

Now

$$\int_0^b e^{-2x} \, dx = \frac{e^{-2x}}{-2} \Big]_0^b = \frac{e^{-2x}}{2} \Big]_b^0$$

$$= \frac{e^{-2(0)}}{2} - \frac{e^{-2b}}{2}$$

$$= \frac{1}{2}(1 - e^{-2b})$$

Then

$$\int_0^{+\infty} e^{-2x}\,dx = \lim_{b \to +\infty} \frac{1}{2}(1 - e^{-2b}) = \frac{1}{2}(1 - 0) = \frac{1}{2}$$

15. $\int_2^{+\infty} \frac{1}{\sqrt{x-1}}\,dx = \lim_{b \to +\infty} \int_2^b \frac{1}{\sqrt{x-1}}\,dx$

To determine the indefinite integral we use the method of substitution. Let $u = x - 1$ so that $du = dx$. Now

$$\int \frac{1}{\sqrt{x-1}}\,dx = \int \frac{1}{\sqrt{u}}\,du = \int u^{-1/2}\,du$$

$$= 2\sqrt{u} + C = 2\sqrt{x-1} + C$$

so that

$$\int_2^b \frac{1}{\sqrt{x-1}}\,dx = 2\sqrt{b-1} - 2\sqrt{2-1}$$

$$= 2(\sqrt{b-1} - 1)$$

Then

$$\int_2^{+\infty} \frac{1}{\sqrt{x-1}}\,dx = \lim_{b \to +\infty} 2(\sqrt{b-1} - 1)$$

Since the limit on the right does not exist we conclude that the improper integral diverges.

17. $\int_0^{+\infty} \frac{x}{(x^2+1)^{1/2}}\,dx = \lim_{b \to +\infty} \int_0^b \frac{x}{(x^2+1)^{1/2}}\,dx$

To compute the indefinite integral we use the method of substitution. Let $u = x^2 + 1$ so that

$$du = 2x\,dx \text{ and } x\,dx = \frac{1}{2}du$$

Now

$$\int \frac{x}{(x^2+1)^{1/2}}\,dx = \int \frac{1}{u^{1/2}}\left(\frac{1}{2}du\right)$$

$$= \int \frac{1}{2}u^{-1/2}\,du = u^{1/2} + C = \sqrt{x^2+1} + C$$

so that

$$\int_0^b \frac{x}{(x^2+1)^{1/2}}\,dx = \sqrt{x^2+1}\Big]_0^b$$

$$= \sqrt{b^2+1} - \sqrt{(0)^2+1}$$

$$= \sqrt{b^2+1} - 1$$

Then

$$\int_0^{+\infty} \frac{x}{(x^2+1)^{1/2}}\,dx = \lim_{b\to+\infty}[\sqrt{b^2+1}-1]$$

Since the limit on the right does not exist we conclude that the improper integral diverges.

21. The total number of trees produced during the lifetime of the forest is

$$\int_0^{+\infty} N'(t)\,dt = \int_0^{+\infty} \frac{2t}{(t^2+4)^2}\,dt = \lim_{b\to+\infty}\int_0^b \frac{2t}{(t^2+4)^2}\,dt$$

We determine the indefinite integral using the method of substitution. Let $u = t^2 + 4$ so that $du = 2t\,dt$. Now

$$\int \frac{2t}{(t^2+4)^2}\,dt = \int \frac{1}{u^2}\,du = \int u^{-2}\,du$$

$$= \frac{u^{-1}}{-1} + C = -\frac{1}{u} + C = -\frac{1}{t^2+4} + C$$

so that

$$\int_0^b \frac{2t}{(t^2+4)^2} dt = \frac{-1}{t^2+4}\Big]_0^b = \frac{1}{t^2+4}\Big]_b^0$$
$$= \frac{1}{0^2+4} - \frac{1}{b^2+4}$$
$$= \frac{1}{4} - \frac{1}{b^2+4}$$

Then

$$\int_0^{+\infty} N'(t)\, dt = \lim_{b \to +\infty} [\frac{1}{4} - \frac{1}{b^2+4}] = \frac{1}{4} - 0 = \frac{1}{4}$$

Thus, 1/4 of a million trees are produced during the lifetime of the forest.

23. We treat the perpetuity as a continuous stream of income, producing income at the rate of $1800 per year, forever. The capitalized value of this income stream is

$$P(+\infty) = \int_0^{+\infty} 1800 e^{-0.06t}\, dt = \lim_{b \to +\infty} 1800 \int_0^b e^{-0.06t}\, dt$$

$$= 1800 \lim_{b \to +\infty} \frac{e^{-0.06b}}{-0.06}\Big]_0^b$$

$$= 1800 \lim_{b \to +\infty} [\frac{e^{-0.06b}}{-0.06} - \frac{1}{-0.06}]$$

$$= \frac{1800}{0.06} \lim_{b \to +\infty} [1 - \frac{1}{e^{0.06b}}] = 30{,}000(1 - 0) = \$30{,}000$$

29. $\int_{-\infty}^{-2} \frac{1}{(1-x)^2} dx = \lim_{a \to -\infty} \int_a^{-2} \frac{1}{(1-x)^2} dx$

To determine the indefinite integral we use the method of substitution. Let $u = 1 - x$ so that $du = -dx$ and $dx = -du$. Now

$$\int \frac{1}{(1-x)^2} \, dx = \int \frac{1}{u^2}(-du) = -\int u^{-2} \, du$$

$$= \frac{-u^{-1}}{-1} + C = \frac{1}{u} + C = \frac{1}{1-x} + C$$

so that

$$\int_a^{-2} \frac{1}{(1-x)^2} \, dx = \frac{1}{1-x}\Big]_a^{-2}$$

$$= \frac{1}{1+2} - \frac{1}{1-a} = \frac{1}{3} - \frac{1}{1-a}$$

Thus,

$$\int_{-\infty}^{-2} \frac{1}{(1-x)^2} \, dx = \lim_{a \to -\infty} \int_a^{-2} \frac{1}{(1-x)^2} \, dx$$

$$= \lim_{a \to -\infty} \left[\frac{1}{3} - \frac{1}{1-a}\right] = \frac{1}{3}$$

Exercise Set 7.7 (Page 467)

3. We want to approximate $\int_{-1}^{1} (1 - x^2) \, dx$ with $n = 4$. First we let

$$\Delta x = \frac{b-a}{n} = \frac{1-(-1)}{4} = \frac{2}{4} = \frac{1}{2}$$

Then

$$x_0 = -1 \qquad y_0 = (1 - (-1)^2) = 0$$
$$x_1 = -\frac{1}{2} \qquad y_1 = (1 - (-\frac{1}{2})^2) = \frac{3}{4}$$
$$x_2 = 0 \qquad y_2 = (1 - (0)^2) = 1$$
$$x_3 = \frac{1}{2} \qquad y_3 = (1 - (\frac{1}{2})^2) = \frac{3}{4}$$
$$x_4 = 1 \qquad y_4 = (1 - (1)^2) = 0$$

$$L_4 = \Delta x [y_0 + y_1 + y_2 + y_3] = \frac{1}{2}[0 + \frac{3}{4} + 1 + \frac{3}{4}] = \frac{1}{2}(\frac{10}{4}) = \frac{5}{4} = 1.25$$

$$R_4 = \Delta x [y_1 + y_2 + y_3 + y_4] = \frac{1}{2}[\frac{3}{4} + 1 + \frac{3}{4} + 0] = \frac{1}{2}(\frac{10}{4}) = \frac{5}{4} = 1.25$$

$$T_4 = \frac{\Delta x}{2}[y_0 + 2y_1 + 2y_2 + 2y_3 + y_4] = \frac{1}{2}[0 + 2(\frac{3}{4}) + 2(1) + 2(\frac{3}{4}) + 0]$$
$$= \frac{1}{4}[0 + \frac{6}{4} + 2 + \frac{6}{4} + 0]$$
$$= \frac{1}{4}(\frac{20}{4}) = \frac{5}{4} = 1.25$$

For the midpoint approximation,

$$x_1 = \frac{-1 + (-1/2)}{2} = -\frac{3}{4} \qquad \tilde{y}_1 = (1 - (-\frac{3}{4})^2) = \frac{7}{16}$$

$$x_2 = \frac{-1/2 + 0}{2} = -\frac{1}{4} \qquad \tilde{y}_2 = (1 - (-\frac{1}{4})^2) = \frac{15}{16}$$

$$x_3 = \frac{0 + 1/2}{2} = \frac{1}{4} \qquad \tilde{y}_3 = (1 - (\frac{1}{4})^2) = \frac{15}{16}$$

$$x_4 = \frac{1/2 + 1}{2} = \frac{3}{4} \qquad \tilde{y}_4 = (1 - (\frac{3}{4})^2) = \frac{7}{16}$$

$$M_4 = \Delta x [\tilde{y}_1 + \tilde{y}_2 + \tilde{y}_3 + \tilde{y}_4] = \frac{1}{2}[\frac{7}{16} + \frac{15}{16} + \frac{15}{16} + \frac{7}{16}]$$

$$= \frac{1}{2}(\frac{44}{16}) = \frac{22}{16} = \frac{11}{8} = 1.375$$

7. We want to approximate $\int_{-3}^{3} x^3 \, dx$ with $n = 6$. First we let
$$\Delta x = \frac{b-a}{n} = \frac{3 - (-3)}{6} = \frac{6}{6} = 1$$

Then

$x_0 = -3$	$y_0 = -27$	$x_1 = -2.5$	$\tilde{y}_1 = -15.625$
$x_1 = -2$	$y_1 = -8$	$x_2 = -1.5$	$\tilde{y}_2 = -3.375$
$x_2 = -1$	$y_2 = -1$	$x_3 = -.5$	$\tilde{y}_3 = -0.125$
$x_3 = 0$	$y_3 = 0$	$x_4 = .5$	$\tilde{y}_4 = 0.125$
$x_4 = 1$	$y_4 = 1$	$x_5 = 1.5$	$\tilde{y}_5 = 3.375$
$x_5 = 2$	$y_5 = 8$	$x_6 = 2.5$	$\tilde{y}_6 = 15.625$
$x_6 = 3$	$y_6 = 27$		

7: The Definite Integral   195

$L_6 = 1 [-27 + (-8) + (-1) + (0) + (1) + (8)] = -27$

$R_6 = 1 [(-8) + (-1) + (0) + (1) + (8) + (27)] = 27$

$M_6 = 1 (-15.625) + (-3.375) + (-0.125) + (0.125) + (3.375) + (15.625)] = 0$

$T_6 = \frac{1}{2}[-27 + 2(-8) + 2(-1) + 2(0) + 2(1) + 2(8) + 27] = 0$

15. We want to approximate $\int_2^5 \ln x \, dx$ with n = 6. First we let
$\Delta x = \frac{b-a}{n} = \frac{5-2}{6} = \frac{3}{6} = \frac{1}{2}$

Then

$x_0 = 2$ $y_0 = \ln 2 = 0.693$ $\tilde{x}_1 = 2.25$ $\tilde{y}_1 = 0.811$
$x_1 = 2.5$ $y_1 = \ln 2.5 = 0.916$ $\tilde{x}_2 = 2.75$ $\tilde{y}_2 = 1.012$
$x_2 = 3$ $y_2 = \ln 3 = 1.099$ $\tilde{x}_3 = 3.25$ $\tilde{y}_3 = 1.179$
$x_3 = 3.5$ $y_3 = \ln 3.5 = 1.253$ $\tilde{x}_4 = 3.75$ $\tilde{y}_4 = 1.322$
$x_4 = 4$ $y_4 = \ln 4 = 1.386$ $\tilde{x}_5 = 4.25$ $\tilde{y}_5 = 1.447$
$x_5 = 4.5$ $y_5 = \ln 4.5 = 1.504$ $\tilde{x}_6 = 4.75$ $\tilde{y}_6 = 1.559$
$x_6 = 5$ $y_6 = \ln 5 = 1.609$

$L_6 = \frac{1}{2}[0.693 + 0.916 + 1.099 + 1.253 + 1.386 + 1.504] = \frac{1}{2}(6.851) = 3.43$

$R_6 = \frac{1}{2}[0.916 + 1.099 + 1.253 + 1.386 + 1.504 + 1.609] = \frac{1}{2}(7.767) = 3.88$

$M_6 = \frac{1}{2}[0.811 + 1.012 + 1.179 + 1.322 + 1.447 + 1.559] = \frac{1}{2}(7.330) = 3.67$

$T_6 = \frac{1}{4}[0.693 + 2(0.916) + 2(1.099) + 2(1.253) + 2(1.386) + 2(1.504) + 1.609]$

$= \frac{1}{4}[0.693 + 1.832 + 2.198 + 2.506 + 2.772 + 3.008 + 1.609]$

$= \frac{1}{4}(14.618) = 3.65$

Review Exercises, (Page 469)

5. The desired area is

$$\int_0^4 e^{3x}\,dx = \frac{1}{3}e^{3x}\Big]_0^4 = \frac{1}{3}e^{3(4)} - \frac{1}{3}e^{3(0)}$$
$$= \frac{1}{3}(e^{12} - 1)$$

7. $\int_1^4 dx = x\Big]_1^4 = 4 - 1 = 3$

9. $\int_1^{27}(x^{1/3} - 5)\,dx = [\frac{3}{4}x^{4/3} - 5x]\Big]_1^{27}$

$$= \frac{3}{4}(27)^{4/3} - 5(27) - [\frac{3}{4}(1)^{4/3} - 5(1)]$$

$$= \frac{3}{4}(\sqrt[3]{27})^4 - 135 - [\frac{3}{4}(1) - 5]$$

$$= \frac{3}{4}(81) - 135 - \frac{3}{4} + 5$$

$$= \frac{3}{4}(81 - 1) - 130 = \frac{3}{4}(80) - 130$$

$$= 60 - 130 = -70$$

13. The desired area is

$$\int_0^2 (2-x)\,dx = \left(2x - \frac{x^2}{2}\right)\Big]_0^2$$

$$= (2(2) - \frac{(2)^2}{2}) - (2(0) - \frac{(0)^2}{2})$$

$$= 4 - \frac{4}{2} - 0$$

$$= 2$$

15. $\int_2^4 (3x - x^2 - \frac{1}{x})\,dx = \left(\frac{3x^2}{2} - \frac{x^3}{3} - \ln|x|\right)\Big]_2^4$

$$= (\frac{3(4)^2}{2} + \frac{(4)^3}{3} - \ln|4|) - (\frac{3(2)^2}{2} + \frac{(2)^3}{3} - \ln|2|)$$

$$= \frac{3}{2}(16 - 4) + \frac{1}{3}(64 - 8) + \ln 2 - \ln 4$$

$$= 18 + \frac{56}{3} + \ln(2/4)$$

$$= \frac{110}{3} + \ln(1/2)$$

$$= \frac{110}{3} + \ln 1 - \ln 2 = \frac{110}{3} - \ln 2$$

17. $\int_1^4 (3\sqrt{x} + 2x^2 - 1)\, dx = \int_1^4 (3x^{1/2} + 2x^2 - 1)\, dx$

$$= \left. (3 \cdot \frac{2}{3} x^{3/2} + \frac{2x^3}{3} - x) \right]_1^4$$

$$= \left. (2x^{3/2} + \frac{2x^3}{3} - x) \right]_1^4$$

$$= (2 \cdot 4^{3/2} + \frac{2}{3} \cdot 4^3 - 4) - (2 \cdot 1^{3/2} + \frac{2}{3} \cdot 1^3 - 1)$$

$$= (2 \cdot 8 + \frac{2}{3} \cdot 64 - 4) - (2 + \frac{2}{3} - 1)$$

$$= (12 + \frac{128}{3}) - (1 + \frac{2}{3})$$

$$= 11 + \frac{126}{3} = 11 + 42 = 53$$

23. The region bounded by $y = x^2$ and $y = \sqrt{x}$ is sketched below

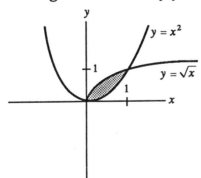

The x-coordinates of the points of intersection are found by solving

$$x^2 = \sqrt{x}$$
$$x^4 = x$$
$$x^4 - x = 0$$
$$x(x^3 - 1) = 0$$
$$x = 0 \quad x^3 = 1$$
$$x = 1$$

Thus we shall integrate over the interval [0,1]. The area is

$$A = \int_0^1 (\sqrt{x} - x^2)\, dx = \frac{2}{3} x^{3/2} - \frac{x^3}{3} \Bigg]_0^1$$

$$= \frac{2}{3}(1)^{3/2} - \frac{(1)^3}{3} - [\frac{2}{3}(0)^{3/2} - \frac{(0)^3}{3}]$$

$$= \frac{2}{3} - \frac{1}{3} - 0 = \frac{1}{3}$$

31. The decrease in reaction time between the fourth and eighth hours of weightlessness is

$$\int_4^8 r'(t)\, dt = \int_4^8 (0.003t^2 - 0.4t)\, dt = \frac{0.03t^3}{3} - \frac{0.4t^2}{2} \Bigg]_4^8$$

$$= 0.001t^3 - 0.2t^2 \Bigg]_4^8$$

$$= (0.001)(8)^3 - 0.2(8)^2 - [0.001(4)^3 - 0.2(4)^2]$$

$$= (0.001)(512) - (0.2)(64) - [0.001(64) - 0.2(16)]$$

$$= 0.512 - 12.8 - 0.064 + 3.2 = -9.152$$

The decrease in reaction time is 9.152 milliseconds.

37. The average value is

$$y_{ave.} = \frac{1}{10-5} \int_5^{10} e^{0.2x}\, dx = \frac{1}{5} [\frac{e^{0.2x}}{0.2}]_5^{10}$$

$$= \frac{1}{5(.2)} [e^{0.2(10)} - e^{0.2(5)}] = e^2 - e = 4.671$$

43. $\int_3^{+\infty} \frac{1}{(x-2)^{1/3}} dx = \lim_{b \to +\infty} \int_3^b \frac{1}{(x-2)^{1/3}} dx$

The indefinite integral is determined using the method of substitution. Let $u = x - 2$ so that $du = dx$ and

$$\int \frac{1}{(x-2)^{1/3}} dx = \int (x-2)^{-1/3} dx = \int u^{-1/3} du$$

$$= \frac{3}{2} u^{2/3} + C = \frac{3}{2} (x-2)^{2/3} + C$$

so that

$$\int_3^b \frac{1}{(x-2)^{1/3}} dx = \frac{3}{2}(x-2)^{2/3} \Big]_3^b$$

$$= \frac{3}{2} [(b-2)^{2/3} - (3-2)^{2/3}]$$

$$= \frac{3}{2} [(b-2)^{2/3} - 1]$$

Then

$$\int_3^{+\infty} \frac{1}{(x-2)^{1/3}} dx = \lim_{b \to +\infty} \int_3^b \frac{3}{2} [(b-2)^{2/3} - 1]$$

Since the limit on the right does not exist we conclude that the improper integral diverges.

55. We want to approximate $\int_0^2 \frac{1}{x+1} dx$ with $n = 4$. First we let

$$\Delta x = \frac{b-a}{n} = \frac{2-0}{4} = \frac{1}{2}$$

Then

$x_0 = 0$ $y_0 = \frac{1}{0+1} = 1$ $x_1 = \frac{1}{4}$ $\tilde{y}_1 = \frac{1}{1/4+1} = \frac{4}{5}$

$$x_1 = \frac{1}{2} \quad y_1 = \frac{1}{1/2+1} = \frac{2}{3} \quad x_2 = \frac{3}{4} \quad \tilde{y}_2 = \frac{1}{3/4+1} = \frac{4}{7}$$

$$x_2 = 1 \quad y_2 = \frac{1}{1+1} = \frac{1}{2} \quad x_3 = \frac{5}{4} \quad \tilde{y}_3 = \frac{1}{5/4+1} = \frac{4}{9}$$

$$x_3 = \frac{3}{2} \quad y_3 = \frac{1}{3/2+1} = \frac{2}{5} \quad x_4 = \frac{7}{4} \quad \tilde{y}_4 = \frac{1}{7/4+1} = \frac{4}{11}$$

$$x_4 = 2 \quad y_4 = \frac{1}{2+1} = \frac{1}{3}$$

$$L_4 = \frac{1}{2}[1 + \frac{2}{3} + \frac{1}{2} + \frac{2}{5}] = \frac{1}{2}(2.5667) = 1.283$$

$$R_4 = \frac{1}{2}[\frac{2}{3} + \frac{1}{2} + \frac{2}{5} + \frac{1}{3}] = \frac{1}{2}(1.9) = 0.950$$

$$M_4 = \frac{1}{2}[\frac{4}{5} + \frac{4}{7} + \frac{4}{9} + \frac{4}{11}] = \frac{1}{2}(4)[\frac{1}{5} + \frac{1}{7} + \frac{1}{9} + \frac{1}{11}] = 2(0.54488) = 1.090$$

$$T_4 = \frac{1}{4}[1 + 2(\frac{2}{3}) + 2(\frac{1}{2}) + 2(\frac{2}{5}) + \frac{1}{3}]$$
$$= \frac{1}{4}[1 + \frac{4}{3} + 1 + \frac{4}{5} + \frac{1}{3}]$$
$$= \frac{1}{4}(4.4666) = 1.117$$

Chapter Test, (Page 472)

1. Since $y = f(x) = \sqrt{4x - 3}$ is positive over the interval [1,3], the area is given by

$$\int_1^3 \sqrt{4x - 3}\, dx$$

The indefinite integral is determined using the method of substitution. Let $u = 4x - 3$ so that $du = 4\, dx$ or $dx = \frac{1}{4} du$. Then

$$\int \sqrt{4x - 3}\, dx = \int (4x - 3)^{1/2}\, dx = \int u^{1/2} (\frac{1}{4} du)$$

$$= \frac{1}{4} \int u^{1/2}\, du = \frac{1}{4} \cdot \frac{2}{3} u^{3/2}$$

$$= \frac{1}{6} u^{3/2} + C = \frac{1}{6}(4x-3)^{3/2} + C$$

Thus

$$\int_1^3 \sqrt{4x-3}\, dx = \frac{1}{6}(4x-3)^{3/2}\Big]_1^3$$

$$= \frac{1}{6}[(4(3)-3)^{3/2} - (4(1)-3)^{3/2}]$$

$$= \frac{1}{6}[9^{3/2} - 1^{3/2}] = \frac{1}{6}[(\sqrt{9})^3 - 1] = \frac{1}{6}(3^3 - 1)$$

$$= \frac{1}{6}(27-1) = \frac{26}{6} = \frac{13}{3}$$

3. Given that f(x) is a continuous function on the interval [0,5], and $\int_0^5 f(x)\, dx = 10$ while $\int_2^5 f(x)\, dx = 4$

(a) $\int_5^2 f(x)\, dx = -\int_2^5 f(x)\, dx = -4$

(b) Since

$$\int_0^2 f(x)\, dx + \int_2^5 f(x)\, dx = \int_0^5 f(x)\, dx$$

we have

$$\int_0^2 f(x)\, dx + 4 = 10$$

or

$$\int_0^2 f(x)\, dx = 10 - 4 = 6$$

(c) $\int_2^2 f(x)\, dx = 0$

5. The change in the amount of gold sold between t = 0 and t = 4 weeks is

$$\int_0^4 A'(t)\, dt = \int_0^4 50\, e^{-3t}\, dt = 50\int_0^4 e^{-3t}\, dt$$

$$= 50\left[\frac{e^{-3t}}{-3}\right]_0^4$$

$$= -\frac{50}{3}[e^{-3(4)} - e^0]$$

$$= +\frac{50}{3}[1 - e^{-12}] = 16.667 \text{ million ounces}$$

7. (a) $\int_1^{+\infty} \frac{1}{x^5} dx = \lim_{b \to +\infty} \int_1^b \frac{1}{x^5} dx$

Now

$$\int \frac{1}{x^5} dx = \int x^{-5} dx = \frac{x^{-4}}{-4} + C$$

so that

$$\int_1^b \frac{1}{x^5} dx = -\frac{1}{4x^4}\Big]_1^b = -\frac{1}{4}[\frac{1}{b^4} - 1]$$

Then

$$\int_1^{+\infty} \frac{1}{x^5} dx = \lim_{b \to +\infty} \int_1^b \frac{1}{x^5} dx = \lim_{b \to +\infty} -\frac{1}{4}[\frac{1}{b^4} - 1] = -\frac{1}{4}[0 - 1] = \frac{1}{4}$$

(b) $\int_2^{\infty} \frac{x}{x^2 - 1} dx = \lim_{b \to \infty} \int_2^b \frac{x}{x^2 - 1} dx$

The indefinite integral is determined using the method of substitution. Let $u = x^2 - 1$ so that $du = 2x\, dx$ or $x\, dx = \frac{1}{2} du$. Then

$$\int \frac{x}{x^2 - 1} dx = \int \frac{1/2\, du}{u} = \frac{1}{2}\int \frac{du}{u} = \frac{1}{2}\ln|u| + C = \frac{1}{2}\ln|x^2 - 1| + C$$

so that

$$\int_2^b \frac{x}{x^2 - 1} dx = \frac{1}{2}[\ln|x^2 - 1|]_2^b$$

$$= \frac{1}{2}[\ln |b^2-1| - \ln |2^2-1|]$$

Thus

$$\int_2^\infty \frac{x}{x^2-1}\,dx = \lim_{b\to\infty} \frac{1}{2}[\ln |b^2-1| - \ln 3]$$

Since the limit on the right does not exist, we conclude that the improper integral diverges.

9. We first find the equilibrium price by equating $S(x)$ and $D(x)$.

$$4 + \frac{x^2}{3} = 22 - \frac{x^2}{6}$$

Multiplying by 6 we have

$$24 + 2x^2 = 132 - x^2$$
$$3x^2 = 108$$
$$x^2 = 36$$
$$x = \pm 6$$

Since the number of units cannot be negative, we have $\tilde{x} = x_E = 6$. Now, using either the supply or the demand function, we obtain $P_E = 16$. Thus

$$CS = \int_0^6 [22 - \frac{x^2}{6} - 16]\,dx = \int_0^6 (6 - \frac{x^2}{6})\,dx = 6x - \frac{x^3}{18}\Big]_0^6$$

$$= [6(6) - \frac{(6)^3}{18}] - [6(0) - \frac{(0)^3}{18}] = 36 - \frac{216}{18} - 0 = 36 - 12 = 24$$

and

$$PS = \int_0^6 [16 - (4 + \frac{x^2}{3})]\,dx = \int_0^6 (12 - \frac{x^2}{3})\,dx = 12x - \frac{x^3}{9}\Big]_0^6$$

$$= [12(6) - \frac{(6)^3}{9}] - [12(0) - \frac{(0)^3}{9}] = 72 - \frac{216}{9} - 0 = 72 - 24 = 48$$

8 Functions of Several Variables

Key Ideas for Review

* A function of the independent variables x and y is a rule, or formula, that determines exactly one value of the dependent variable z for each ordered pair of values (x, y) for which the rule is defined. We denote this value of z by f(x, y) and write

$$z = f(x, y)$$

* A three dimensional Cartesian coordinate system sets up a one-to-one correspondence between points in three-dimensional space and ordered triples of real numbers.

* The graph of a function f of x and y, called a surface, is the set of all points, (x, y, z) in three-dimensional space for which z = f(x, y).

* The graph of a first degree equation in x, y, and z is a plane.

* If z = f(x, y), then the partial derivative of f with respect to x--denoted by f_x, $\partial f/\partial x$, or $\partial z/\partial x$--is obtained by taking the ordinary derivative of f with respect to x, while the variable y remains constant. Similarly, the partial derivative of f with respect to y--denoted by f_y, $\partial f/\partial y$, or $\partial z/\partial y$--is obtained by taking the ordinary derivative of f with respect to y, while the variable x remains constant.

* If z = f(x, y), then the second-order partial derivatives are defined by

$$\frac{\partial}{\partial x}(\frac{\partial z}{\partial x}) = \frac{\partial^2 z}{\partial x^2} \qquad \text{or} \qquad f_{xx}$$

$$\frac{\partial}{\partial y}(\frac{\partial z}{\partial y}) = \frac{\partial^2 z}{\partial y^2} \qquad \text{or} \qquad f_{yy}$$

$$\frac{\partial}{\partial x}(\frac{\partial z}{\partial y}) = \frac{\partial^2 z}{\partial x \partial y} \qquad \text{or} \qquad f_{yx}$$

$$\frac{\partial}{\partial y}\left(\frac{\partial z}{\partial x}\right) = \frac{\partial^2 z}{\partial y \partial x} \quad \text{or} \quad f_{xy}$$

(the mixed partials f_{xy} and f_{yx} are equal whenever f_x, f_y, f_{xy}, and f_{yx} are continuous)

* The notion of partial derivative extends to functions of three or more variables.

* $f(x, y)$ has a relative maximum (or relative minimum) at (a, b) if there is a circular disk in the xy-plane centered at (a, b) throughout which f is defined, such that for all points (x, y) in this disk $f(x, y) \geq f(a, b)$ (or $f(a, b) \leq f(x, y)$).

* (a, b) is a critical point of f if (a, b) is the center of a circular disk throughout which f is defined and either $f_x(a, b) = 0$ and $f_y(a, b) = 0$, or $f_x(a, b)$ and $f_y(a, b)$ do not both exist.

* If f has a relative extremum at (a, b), then (a, b) is a critical point of f. Critical points are candidates for points where f has relative extrema.

* If (a, b) is a critical point of f, then f may have a relative maximum, a relative minimum, or no relative extremum at (a, b). A critical point of f at which f does not have a relative extremum is called a saddle point.

* Second partials test for relative extrema: Let (a, b) be a critical point of f and let

    $$M = f_{xx}(a, b) \cdot f_{yy}(a, b) - [f_{xy}(a, b)]^2.$$

    (a) If $M > 0$ and $f_{xx}(a, b) < 0$, then f has a relative maximum at (a, b).

    (b) If $M > 0$ and $f_{xx}(a, b) > 0$, then f has a relative minimum at (a, b).

    (c) If $M < 0$ then f has no relative extremum at (a, b).

    (d) If $M = 0$, then f may or may not have a relative extremum at (a, b).

* The method of Lagrange multipliers: To maximize or minimize $f(x, y)$ subject to the constraint $g(x, y) = 0$, proceed as follows:

    Step 1: Form $F(x, y, \lambda) = f(x, y) + \lambda g(x, y)$.

    Step 2: Solve the system

$$F_x(x, y, \lambda) = 0$$

$$F_y(x, y, \lambda) = 0$$

$$F_\lambda(x, y, \lambda) = 0$$

Step 3: For each solution (a, b, λ_0) to the system obtained in step 2, evaluate $f(a, b)$. The constrained maximum or minimum value of f, if it exists, will be among these values.

* The method of Lagrange multipliers extends to the problem of maximizing or minimizing a function $f(x, y, z)$ subject to the constraint $g(x, y, z) = 0$.

* The total differential of a function $z = f(x, y)$ is defined by

$$dz = \frac{\partial z}{\partial x} dx + \frac{\partial z}{\partial y} dy$$

This definition extends to functions of three or more variables.

* $df \approx \Delta f$.

* A region R of Type I in the xy-plane is the set of all points (x, y) such that $a \le x \le b$, and $g_1(x) \le y \le g_2(x)$, where g_1 and g_2 are continuous functions over the interval $[a, b]$.

* A region R of Type II in the xy-plane is the set of all points (x, y) such that $c \le y \le d$ and $h_1(y) \le x \le h_2(y)$, where h_1 and h_2 are continuous functions over the interval $[c, d]$.

* Suppose f is a function of x and y that is continuous over a region R of Type I or Type II. Partition R into n rectangles whose sides are parallel to the coordinate axes, and each having area $\Delta x \Delta y$. In each rectangle R_i that is entirely contained in R choose any arbitrary point (x_i, y_i) and form the Reimann sum

$$f(x_1, y_1) \Delta x \Delta y + f(x_2, y_2) \Delta x \Delta y + \ldots + f(x_n, y_n) \Delta x \Delta y$$

Then, the limit of the Reimann sums, as $n \rightarrow +\infty$ and Δx and Δy approach zero, exists and is independent of the choice of points (x_i, y_i). This limit is called the double integral of f over R and is denoted by

8: Functions of Several Variables.. 207

$$\iint_R f(x, y)\, dy\, dx$$

* An iterated integral is an expression of the form

$$\int_a^b \left(\int_{g_1(x)}^{g_2(x)} f(x, y)\, dy \right) dx$$

or

$$\int_c^d \left(\int_{h_1(x)}^{h_2(x)} f(x, y)\, dx \right) dy$$

where g_1 and g_2 are continuous over [a, b] and h_1 and h_2 are continuous over [c, d].

* If f is a function of x and y which is continuous over a region R of Type I, then

$$\iint_R f(x, y)\, dy\, dx = \int_a^b \left(\int_{g_1(x)}^{g_2(x)} f(x, y)\, dy \right) dx$$

* If f is a function of x and y which is continuous over a region R of Type II, then

$$\iint_R f(x, y)\, dy\, dx = \int_c^d \left(\int_{h_1(x)}^{h_2(x)} f(x, y)\, dx \right) dy$$

Exercise Set 8.1, (Page 481)

1. Since $f(x, y) = 5 + 3x - 2y$

    (a) $f(1, 2) = 5 + 3(1) - 2(2) = 8 - 4 = 4$

    (b) $f(-2, 3) = 5 + 3(-2) - 2(3) = 5 - 6 - 6 = -7$

    (c) $f(0, 4) = 5 + 3(0) - 2(4) = 5 - 8 = -3$

    (d) $f(-1, -3) = 5 + 3(-1) - 2(-3) = 5 - 3 + 6 = 8$

    (e) The function f(x, y) is defined for all (x, y). Thus the domain is all points (x, y).

5. Since $f(x, y, z) = x^2 + y^3 - xz + z^2$

    (a) $f(1, -1, 2) = (1)^2 + (-1)^3 - (1)(2) + (2)^2 = 1 - 1 - 2 + 4 = 2$

(b) $f(0, 0, 4) = (0)^2 + (0)^3 - (0)(4) + (4)^2 = 16$

(c) $f(-2, 3, 1) = (-2)^2 + (3)^3 - (-2)(1) + (1)^2 = 4 + 27 + 2 + 1 = 34$

(d) $f(2, 1, -2) = (2)^2 + (1)^3 - (2)(-2) + (-2)^2 = 4 + 1 + 4 + 4 = 13$

(e) The function $f(x, y, z)$ is defined for all (x, y, z). Thus the domain is all points (x, y, z).

9. Since $K(a, b, c) = ce^{a} + b^2$

   (a) $K(2, 3, 1) = (1)e^2 + (3)^2 = e^{11}$

   (b) $K(1, 0, 3) = 3e^1 + (0)^2 = 3e$

11. (a) To determine $f(2 + h, 3)$ we replace each occurrence of x in the rule for $f(x, y)$ by $2 + h$ and each occurrence of y by 3. Thus, when

    $$f(x, y) = 2x + 3y$$

    we have

    $$f(2 + h, 3) = 2(2 + h) + 3(3)$$
    $$= 4 + 2h + 9 = 13 + 2h$$

    (b) To determine $f(x + h, y)$ we replace each occurrence of x in the rule for $f(x, y)$ by $x + h$. Thus,

    $$f(x + h, y) = 2(x + h) + 3y = 2x + 2h + 3y$$

    (c) Since $f(2, 3) = 2(2) + 3(3) = 4 + 9 = 13$ and

    $$f(2 + h, 3) = 13 + 2h,$$

    we have

    $$f(2 + h, 3) - f(2, 3) = 13 + 2h - 13 = 2h$$

    (d) Since $f(x, y) = 2x + 3y$ and $f(x + h, y) = 2x + 2h + 3y$, we have

    $$f(x + h, y) - f(x, y) = (2x + 2h + 3y) - (2x + 3y) = 2h$$

    (e) Using the result of (d) we have

$$\frac{f(x+h, y) - f(x, y)}{h} = \frac{2h}{h} = 2$$

17. (a) C(3, 12) = 10 + 2(3)² + 3 (12) = 10 + 18 + 36 = 64. Thus, the manufacturer's cost is $64,000 when there are 3 employees and $12,000 is spent on materials.

    (b) C(15, 8) = 10 + 2 (15)² + 3(8) = 10 + 450 + 24 = 484. Thus, the manufacturer's cost is $484,000 when there are 15 employees and $8000 is spent on materials.

Exercise Set 8.2, (Page 491)

1. To find f_x we treat y as a constant and differentiate f with respect to x, obtaining

$$f_x = \frac{\partial}{\partial x}[x^3 + xy + 3y^2]$$

$$= \frac{\partial}{\partial x}[x^3] + y\frac{\partial}{\partial x}[x] + \frac{\partial}{\partial x}[3y^2]$$

$$= 3x^2 + y(1) + 0$$

$$= 3x^2 + y$$

To determine f_y we treat x as a constant and differentiate f with respect to y, obtaining

$$f_y = \frac{\partial}{\partial y}[x^3 + xy + 3y^2]$$

$$= \frac{\partial}{\partial y}[x^3] + x\frac{\partial}{\partial y}[y] + \frac{\partial}{\partial y}[3y^2]$$

$$= 0 + x(1) + 6y$$

$$= x + 6y$$

Now

$$f_{xy} = \frac{\partial}{\partial y}[f_x] = \frac{\partial}{\partial y}[3x^2 + y]$$

$$= \frac{\partial}{\partial y}[3x^2] + \frac{\partial}{\partial y}[y]$$

$$= 0 + 1 = 1$$

7. To determine f_x we treat y as a constant and differentiate f with respect to x, obtaining

$$f_x = \frac{\partial}{\partial x}[e^{xy^2}] = e^{xy^2} \frac{\partial}{\partial x}[xy^2]$$

$$= e^{xy^2} \cdot y^2 = y^2 e^{xy^2}$$

To determine f_y we treat x as a constant and differentiate f with respect to y, obtaining

$$f_y = \frac{\partial}{\partial y}[e^{xy^2}] = e^{xy^2} \frac{\partial}{\partial y}[xy^2]$$

$$= e^{xy^2}(2xy) = 2xy e^{xy^2}$$

Now

$$f_{xy} = \frac{\partial}{\partial y}[f_x] = e^{xy^2} \frac{\partial}{\partial y}[y^2 e^{xy^2}]$$

The product rule yields

$$f_{xy} = y^2 \frac{\partial}{\partial y}[e^{xy^2}] + e^{xy^2} \frac{\partial}{\partial y}[y^2]$$

$$= y^2 e^{xy^2}(2xy) + e^{xy^2}(2y)$$

$$= 2y e^{xy^2}(xy^2 + 1)$$

15. Now

$$f_x(x, y) = \frac{\partial}{\partial x}[xe^y + y^2 e^x]$$

$$= e^y \frac{\partial}{\partial x}[x] + y^2 \frac{\partial}{\partial x}[e^x]$$

$$= e^y + y^2 e^x,$$

$$f_y(x,y) = \frac{\partial}{\partial y}[xe^y + y^2 e^x]$$

$$= x\frac{\partial}{\partial y}[e^y] + e^x \frac{\partial}{\partial y}[y^2]$$

$$= xe^y + 2ye^x$$

and

$$f_{yx}(x,y) = \frac{\partial}{\partial x}[f_y(x,y)]$$

$$= \frac{\partial}{\partial x}[xe^y + 2ye^x]$$

$$= e^y \frac{\partial}{\partial x}[x] + 2y \frac{\partial}{dx}[e^x]$$

$$= e^y + 2ye^x$$

Thus,

$$f_x(0,1) = e^1 + (1)^2 e^0 = e + 1$$

$$f_y(0,1) = 0 \cdot e^1 + 2(1)e^0 = 2,$$

and

$$f_{yx}(0,1) = e^1 + 2(1)e^0 = e + 2$$

17. First, we form the partial derivatives

$$f_x = \frac{\partial}{dx}[2x^2 + xy - y^3] = 4x + y$$

and

$$f_y = \frac{\partial}{dy}[2x^2 + xy - y^3] = x - 3y^2$$

Then, we obtain the second-order partial derivatives

$$f_{xx} = \frac{\partial}{\partial x}[f_x] = \frac{\partial}{\partial x}[4x + y] = 4$$

$$f_{yy} = \frac{\partial}{\partial y}[f_y] = \frac{\partial}{\partial y}[x - 3y^2] = -6y$$

$$f_{xy} = \frac{\partial}{\partial y}[f_x] = \frac{\partial}{\partial y}[4x + y] = 1$$

$$f_{yx} = \frac{\partial}{\partial x}[f_y] = \frac{\partial}{\partial x}[x - 3y^2] = 1$$

25. We have

$$f_x(x, y) = \frac{\partial}{\partial x}(5x^2y - y^3) = 5y\frac{\partial}{\partial x}[x^2] - \frac{\partial}{\partial x}[y^3]$$

$$= 5y(2x) - 0 = 10xy$$

$$f_y(x, y) = \frac{\partial}{\partial y}(5x^2y - y^3) = 5x^2\frac{\partial}{\partial y}(y) - \frac{\partial}{\partial y}(y^3)$$

$$= 5x^2 - 3y^2$$

$$f_{xx}(x, y) = \frac{\partial}{\partial x}[10xy] = 10y\frac{\partial}{\partial x}[x] = 10y$$

$$f_{xy}(x, y) = \frac{\partial}{\partial x}[10xy] = 10x\frac{\partial}{\partial y}(y) = 10x$$

$$f_{yx}(x, y) = \frac{\partial}{\partial x}(5x^2 - 3y^2) = \frac{\partial}{\partial x}(5x^2) - \frac{\partial}{\partial x}(3y^2)$$

$$= 10x - 0 = 10x$$

$$f_{yy}(x, y) = \frac{\partial}{\partial y}(5x^2 - 3y^2) = \frac{\partial}{\partial y}(5x^2) - \frac{\partial}{\partial y}(3y^2)$$

$$= 0 - 6y = -6y$$

Thus

$$f_{xx}(-3, 2) = 10(2) = 20$$
$$f_{xy}(-3, 2) = 10(-3) = -30$$
$$f_{yx}(-3, 2) = 10(-3) = -30$$
$$f_{yy}(-3, 2) = -6(2) = -12$$

29. We have

$$f_x(x, y, z) = \frac{\partial}{\partial x}(xy - yz + xz)$$

$$= y\frac{\partial}{\partial x}[x] - \frac{\partial}{\partial x}[yz] + z\frac{\partial}{\partial x}[x]$$

$$= y - 0 + z = y + z$$

$$f_y(x, y, z) = \frac{\partial}{\partial y}(xy - yz + xz)$$

$$= x\frac{\partial}{\partial y}[y] - z\frac{\partial}{\partial y}[y] + \frac{\partial}{\partial y}[xz]$$

$$= x - z + 0 = x - z$$

$$f_z(x, y, z) = \frac{\partial}{\partial z}(xy - yz + xz)$$

$$= \frac{\partial}{\partial z}(xy) - y\frac{\partial}{\partial z}[z] + x\frac{\partial}{\partial z}[z]$$

$$= 0 - y + x = x - y$$

Thus

$$f_x(3, -4, 2) = -4 + 2 = -2$$
$$f_y(3, -4, 2) = 3 - 2 = 1$$
$$f_z(3, -4, 2) = 3 - (-4) = 7$$

35. (a) $\frac{\partial Q}{\partial x} = \frac{\partial}{\partial x}[60x^{2/3}y^{1/3}] = 60y^{1/3}\frac{\partial}{\partial x}[x^{2/3}]$

$$= 60y^{1/3} \cdot \frac{2}{3}x^{-1/3} = 40x^{-1/3}y^{1/3}$$

and

$$\frac{\partial Q}{\partial y} = \frac{\partial}{\partial y}[60x^{2/3}y^{1/3}] = 60x^{2/3}\frac{\partial}{\partial y}[y^{1/3}]$$

$$= 60x^{2/3} \cdot (1/3)y^{-2/3} = 20x^{2/3}y^{-2/3}$$

(b) $\left.\dfrac{dQ}{dx}\right|_{(8,27)} = 40(8)^{-1/3}(27)^{1/3} = 40(1/2)(3) = 60$

and

$\left.\dfrac{dQ}{dy}\right|_{(8,27)} = 20(8)^{2/3}(27)^{-2/3} = 20(4)(1/9) = \dfrac{80}{9}$

(c) When 8 units of labor and 27 units of capital are used, an increase of one unit of labor produces approximately 60 additional cameras and an increase of one unit of capital produces approximately 9 additional cameras.

39. (a) The number of additional roofs that can be completed per year if the contractor hires one more roofer is approximately $N_x(20, 12)$ and

$N_x = \dfrac{\partial}{dx}[20x + 3xy + y^2]$

$= 20 + 3y$

Thus,

$N_x(20, 12) = 20 + 3(12) = 56$

and approximately 56 additional roofs can be completed.

(b) The number of additional roofs that can be completed per year if the contractor hires one more apprentice is approximately $N_y(20, 12)$ and

$N_y = \dfrac{\partial}{dy}[20x + 3xy + y^2]$

$= 3x + 2y$

Thus,

$N_y(20, 12) = 3(20) + 2(12) = 84$

and approximately 84 additional roofs can be completed.

Exercise Set 8.3, (Page 507)

1. To find the critical points, we determine f_x and f_y and set them equal to zero. In this case

    $$f_x = 2x - 6$$
    $$f_y = 2y + 3$$

    Thus, we must solve the system of equations

    $$2x - 6 = 0$$
    $$2y + 3 = 0$$

    The only solution of this system is (3, -3/2). Thus (3, -3/2) is the only critical point of f.

7. To find the critical points, we determine f_x and f_y and set them equal to zero. In this case,

    $$f_x = 6x^2 - 6x$$
    $$f_y = -2y$$

    Thus we must solve the system of equations

    $$6x^2 - 6x = 0$$
    $$-2y = 0$$

    The points (1, 0) and (0, 0) are the only solutions of this system and consequently are the only critical points of f.

9. Since

    $$f(x, y) = x^2 + xy + y^2 - 4x - 5y,$$

    we have

    $$f_x(x, y) = 2x + y - 4,$$
    $$f_y(x, y) = x + 2y - 5,$$
    $$f_{xx}(x, y) = 2,$$
    $$f_{yy}(x, y) = 2,$$

    and

    $$f_{xy}(x, y) = 1$$

To find the critical points, we must solve the system of equations

$$2x + y - 4 = 0$$
$$x + 2y - 5 = 0$$

We solve the first equation for y in terms of x, obtaining

$$y = 4 - 2x$$

and substitute this value of y in the second equation, obtaining

$$x + 2(4 - 2x) - 5 = 0$$

or

$$x + 8 - 4x - 5 = 0$$
$$-3x + 3 = 0$$

or

$$x = 1$$

The corresponding value of y is

$$y = 4 - 2(1) = 2$$

Hence, the only critical point of f is (1, 2). We find that

$$f_{xx}(1, 2) = 2, \ f_{yy}(1,2) = 2, \text{ and } f_{xy}(1, 2) = 1$$

Next

$$M = f_{xx}(1, 2) f_{yy}(1, 2) - [f_{xy}(1, 2)]^2$$
$$= (2)(2) - (1)^2 = 3$$

Since $M > 0$ and $f_{xx}(1, 2) > 0$, it follows from the second partials test that f has a relative minimum at (1, 2). The relative minimum is $f(1, 2) = -7$.

13. Since

$$f(x,y) = x^3 + xy - y^3,$$

we have

$$f_x(x, y) = 3x^2 + y,$$
$$f_y(x, y) = x - 3y^2,$$
$$f_{xx}(x, y) = 6x,$$
$$f_{yy}(x, y) = -6y,$$

and

$$f_{xy}(x, y) = 1$$

To find the critical points of f, we must solve the system of equations

$$3x^2 + y = 0$$
$$x - 3y^2 = 0$$

We solve the first equation for y, obtaining

$$y = -3x^2$$

and substitute this value of y into the second equation, obtaining

$$x - 3(-3x^2)^2 = 0$$

or

$$x - 27x^4 = 0$$
$$x(1 - 27x^3) = 0$$
$$x = 0 \text{ and } x = 1/3$$

The corresponding values of y are

$$y = -3(0)^2 = 0$$

and

$$y = -3(1/3)^2 = -1/3$$

Thus, (0, 0) and (1/3, -1/3) are critical points of f. We find that

$$f_{xx}(0, 0) = 0, f_{yy}(0, 0) = 0, \text{ and } f_{xy}(0, 0) = 1$$

so that

$$M = f_{xx}(0, 0) f_{yy}(0, 0) - [f_{xy}(0, 0)]^2$$
$$= (0)(0) - (1)^2 = -1$$

Since M < 0 we conclude from the second partials test that f has no extremum at (0, 0). We also have

$$f_{xx}(1/3, -1/3) = 2, f_{yy}(1/3, -1/3) = 2 \text{ and } f_{xy}(1/3, -1/3) = 1$$

so that

$$M = f_{xx}(1/3, -1/3)\, f_{yy}(1/3, -1/3) - [f_{xy}(1/3, -1/3)]^2$$

$$= (2)(2) - (1)^2 = 3$$

Since $M > 0$ and $f_{xx}(1/3, -1/3) > 0$, we conclude from the second partials test that f has a relative minimum at $(1/3, -1/3)$. The relative minimum is $f(1/3, -1/3) = -1/27$.

15. Since $f(x, y) = 2x^3 - 6xy + 3y^2 + 6x - 18y$, we have

$$f_x(x, y) = 6x^2 - 6y + 6,$$
$$f_y(x, y) = -6x + 6y - 18,$$
$$f_{xx}(x, y) = 12x,$$
$$f_{yy}(x, y) = 6,$$

and

$$f_{xy}(x, y) = -6.$$

To find the critical points of f we must solve the system of equations

$$6x^2 - 6y + 6 = 0$$
$$-6x + 6y - 18 = 0$$

that is,

$$x^2 - y + 1 = 0$$
$$-x + y - 3 = 0$$

We solve the first equation for y obtaining

$$y = x^2 + 1$$

and substitute this value of y into the second equation obtaining

$$x^2 - x - 2 = 0$$
$$(x - 2)(x + 1) = 0$$

x = 2 and x = -1

The corresponding values of y are

$$y = (2)^2 + 1 = 5$$

and

$$y = (-1)^2 + 1 = 2$$

Thus, the points (2, 5) and (-1, 2) are the only critical points of f. We find that

$$f_{xx}(2, 5) = 24, f_{yy}(2, 5) = 6, \text{ and } f_{xy}(2, 5) = -6.$$

so that

$$M = f_{xx}(2, 5) f_{yy}(2, 5) - [f_{xy}(2, 5)]^2$$

$$= (24)(6) - (-6)^2 = 108$$

Since M > 0 and $f_{xx}(2, 5) > 0$, we conclude from the second partials test that f has a relative minimum at (2, 5). The relative minimum is f(2, 5) = -47. We also have

$$f_{xx}(-1, 2) = -12, f_{yy}(-1, 2) = 6, \text{ and } f_{xy}(-1, 2) = -6$$

so that

$$M = f_{xx}(-1,2) f_{yy}(-1, 2) - [f_{xy}(-1,2)]^2$$

$$= (-12)(6) - (-6)^2 = -108$$

Since M < 0 we conclude from the second partials test that f has no extremum at (-1, 2).

21. Since the number of salespeople and the number of stores are positive, we maximize the profit over the region x > 0, y > 0. Now

$$P(x, y) = 120,000 - (60 - x)^2 - (10 - y)^2$$

so that

$$P_x(x, y) = 2(60 - x),$$
$$P_y(x, y) = 2(10 - y),$$
$$P_{xx}(x, y) = -2,$$
$$P_{yy}(x, y) = -2,$$

and

$$P_{xy}(x, y) = 0.$$

It is clear that (60, 10) is the only critical point. Now

$$P_{xx}(60, 10) = -2, P_{yy}(60, 10) = -2, \text{ and } P_{xy}(60, 10) = 0$$

so that

$$M = P_{xx}(60, 10) P_{yy}(60, 10) - [P_{xy}(60, 10)]^2$$

$$= (-2)(-2) - (0)^2 = 4$$

Since $M > 0$ and $P_{xx}(60, 10) < 0$, the profit is maximum at (60, 10). The maximum profit is

$$P(60, 10) = 120{,}000 \text{ dollars.}$$

25. Let x, y, and z denote the dimensions of the aquarium in meters. Since the volume of the aquarium is 32 cubic meters, we have

$$xyz = 32$$

so that

$$z = \frac{32}{xy}$$

The base of the aquarium has area xy, two vertical sides have area xz, and two vertical sides have area yz. The total surface area is

$$S = xy + 2xz + 2yz$$

We substitute the above expression for z and write S as a function of x and y only. Specifically,

$$S(x, y) = xy + 2x\left(\frac{32}{xy}\right) + 2y\left(\frac{32}{xy}\right)$$

$$= xy + \frac{64}{y} + \frac{64}{x}$$

The problem reduces to minimizing S(x, y) over the region $x > 0$, $y > 0$. Now

$$S_x(x, y) = y - \frac{64}{x^2},$$
$$S_y(x, y) = x - \frac{64}{y^2},$$
$$S_{xx}(x, y) = \frac{128}{x^3},$$
$$S_{yy}(x, y) = \frac{128}{y^3},$$

and

$$S_{xy}(x, y) = 1.$$

To find the critical points we must solve the system of equations

$$y - \frac{64}{x^2} = 0$$
$$x - \frac{64}{y^2} = 0$$

We solve the first equation for y obtaining

$$y = \frac{64}{x^2}$$

We substitute this value of y into the second equation obtaining

$$x - \frac{64}{(\frac{64}{x^2})^2} = 0$$

or

$$x - \frac{x^4}{64} = 0$$
$$x(1 - \frac{x^3}{64}) = 0$$
$$x = 0 \text{ and } x = 4$$

Note that $x = 0$ must be excluded since the first equation of the system involves division by x. When $x = 4$, the corresponding value of y is

$$y = \frac{64}{(4)^2} = 4$$

Thus, (4, 4) is the only critical point of S(x, y). Now

$$S_{xx}(4, 4) = 2, S_{yy}(4, 4) = 2, \text{ and } S_{xy}(4, 4) = 1$$

so that

$$M = S_{xx}(4, 4) S_{yy}(4, 4) - [S_{xy}(4, 4)]^2$$

$$= (2)(2) - (1)^2 = 3$$

Since $M > 0$ and $S_{xx}(4, 4) > 0$, S has a relative minimum at (4, 4). The corresponding value of z is

$$z = \frac{32}{(4)(4)} = 2$$

and

$$S(4, 4) = (4)(4) + \frac{64}{4} + \frac{64}{4} = 48$$

The aquarium of least surface area has a 4 meter square base and is 2 meters high. The minimum surface area is 48 square meters.

27. We organize the data in the table

x_i	x_i^2	y_i	$x_i y_i$
2	4	2	4
3	9	3	9
4	16	4	16
5	25	6	30
Sum = 14	Sum = 54	Sum = 15	Sum = 59

Since there are r = 4 data points, the system of equations to be solved for the line of best fit is

$$4b + 14m = 15$$
$$14b + 54m = 59$$

Multiply the first equation by 7 and the second equation by 2 obtaining

8: Functions of Several Variables.. 223

$$28b + 98m = 105$$
$$28b + 108m = 118$$

Subtraction yields

$$10m = 13$$

or

$$m = \frac{13}{10}.$$

Substitution into the original equation yields $b = \frac{-4}{5}$, so the line of best fit is

$$y = \frac{13}{10}x - \frac{4}{5}.$$

Exercise Set 8.4, (Page 517)

3. We first rewrite the constraint as

$$g(x, y) = x^2 + y^2 - 72 = 0$$

and form the new function

$$F(x, y, \lambda) = x + y + \lambda(x^2 + y^2 - 72)$$

We next solve the system

$$F_x(x, y, \lambda) = 1 + 2\lambda x = 0$$
$$F_y(x, y, \lambda) = 1 + 2\lambda y = 0$$
$$F_\lambda(x, y, \lambda) = x^2 + y^2 - 72 = 0$$

Using the first equation, we obtain

$$x = -\frac{1}{2\lambda}$$

The second equation yields

$$y = -\frac{1}{2\lambda}$$

When these are substituted into the third equation we obtain

$$\frac{1}{4\lambda^2} + \frac{1}{4\lambda^2} - 72 = 0$$

or

$$\frac{1}{2\lambda^2} = 72$$

$$\frac{1}{144} = \lambda^2$$

so $\lambda = \pm 1/12$.

When $\lambda = 1/12$

$$x = -\frac{1}{2\lambda} = -\frac{1}{2(1/12)} = -6$$

and

$$y = -\frac{1}{2\lambda} = -\frac{1}{2(1/12)} = -6$$

When $\lambda = -1/12$,

$$x = -\frac{1}{2\lambda} = -\frac{1}{2(-1/12)} = 6$$

and

$$y = -\frac{1}{2\lambda} = -\frac{1}{2(-1/12)} = 6$$

Thus (-6, -6, 1/12) and (6, 6, -1/12) are solutions of the system of equations. Now since $f(x, y) = x + y$,

$$f(-6, -6) = -6 - 6 = -12$$

and

$$f(6, 6) = 6 + 6 = 12$$

Hence the constrained maximum value of 12 occurs at (6, 6) and the constrained minimum value of -12 occurs at (-6, -6).

7. We first rewrite the constraint as

$$g(x, y, z) = x + y + 2z - 1 = 0$$

Next, form the new function

$$F(x, y, z, \lambda) = x^2 + y^2 + z^2 + \lambda(x + y + 2z - 1)$$

Now, solve the system

$$F_x(x, y, z, \lambda) = 2x + \lambda = 0$$
$$F_y(x, y, z, \lambda) = 2y + \lambda = 0$$
$$F_z(x, y, z, \lambda) = 2z + 2\lambda = 0$$
$$F_\lambda(x, y, z, \lambda) = x + y + 2z - 1 = 0$$

The first three equations of this system reduce to

$$x = -\lambda/2$$
$$y = -\lambda/2$$
$$z = -\lambda$$

Substitution into the fourth equation yields

$$-\frac{\lambda}{2} - \frac{\lambda}{2} + 2(-\lambda) - 1 = 0$$

or

$$-3\lambda - 1 = 0$$
$$\lambda = -1/3$$

When $\lambda = -1/3$, we have

$$x = -\lambda/2 = 1/6$$
$$y = -\lambda/2 = 1/6$$

and

$$z = -\lambda = 1/3$$

Thus, $(1/6, 1/6, 1/3, -1/3)$ is the only solution of the system of equations and consequently the constrained minimum value occurs at $(1/6, 1/6, 1/3)$. Since

$$f(1/6, 1/6, 1/3) = (1/6)^2 + (1/6)^2 + (1/3)^2$$

$$= \frac{1}{36} + \frac{1}{36} + \frac{1}{9} = \frac{1+1+4}{36} = \frac{1}{6}$$

the constrained minimum value is 1/6.

11. Let x and y denote the dimensions of the field in feet. The total amount of fencing used is 5x + 2y feet. The area to be maximized is given by

$$A(x, y) = xy$$

Thus, we maximize A(x, y) subject to the constraint 5x + 2y = 3600. We first rewrite the constraint in the form

$$g(x, y) = 5x + 2y - 3600 = 0$$

Next, form the function

$$F(x, y, \lambda) = A(x, y) + \lambda g(x, y)$$
$$= xy + \lambda(5x + 2y - 3600)$$

Now, solve the system of equations

$$F_x(x, y, \lambda) = y + 5\lambda = 0$$
$$F_y(x, y, \lambda) = x + 2\lambda = 0$$
$$F_\lambda(x, y, \lambda) = 5x + 2y - 3600 = 0$$

The first two equations reduce to

$$y = -5\lambda$$

and

$$x = -2\lambda$$

When these values are substiuted into the third equation, we obtain

$$5(-2\lambda) + 2(-5\lambda) - 3600 = 0$$
$$-20\lambda - 3600 = 0$$

or

$$\lambda = -180$$

When $\lambda = -180$,

$$y = -5\lambda = 900$$

and

$$x = -2\lambda = 360$$

Thus, (360, 900, -180) is the only solution of the system of equations. Hence A has a constrained maximum at (360, 900) and

$$A(360, 900) = 324{,}000$$

so that the constrained maximum value is 324,000 square feet.

15. Let x, y, and z denote the three positive numbers. We must maximize $f(x, y, z) = xyz$ subject to the constraint $x + y + z = 30$. We first rewrite the constraint as

$$g(x, y, z) = x + y + z - 30 = 0$$

Next, form the function

$$\begin{aligned}F(x, y, z, \lambda) &= f(x, y, z) + \lambda g(x, y, z) \\ &= xyz + \lambda(x + y + z - 30)\end{aligned}$$

Now, solve the system of equations

$$\begin{aligned}F_x(x, y, z, \lambda) &= yz + \lambda = 0 \\ F_y(x, y, z, \lambda) &= xz + \lambda = 0 \\ F_z(x, y, z, \lambda) &= xy + \lambda = 0 \\ F_\lambda(x, y, z, \lambda) &= x + y + z - 30 = 0\end{aligned}$$

Subtracting the second equation from the first equation yields

$$z(y - x) = 0$$

Then,

$$z = 0 \quad \text{or} \quad y = x$$

We exclude $z = 0$ from consideration since it is not a positive number. Subtracting the third equation from the second equation yields

$$x(z - y) = 0$$

Then

$$x = 0 \quad \text{or} \quad z = y$$

We also exclude $x = 0$ from consideration since it is not a positive number. Substituting $x = y$ and $z = y$ into the fourth equation yields

$$y + y + y - 30 = 0$$
$$y = 10$$

so that $x = z = 10$. Substituting these results into the third equation yields

$$xy + \lambda = 0$$
$$(10)(10) + \lambda = 0$$

or

$$\lambda = -100$$

Thus, $(10, 10, 10, -100)$ is the only solution of the system of equations with positive x-, y-, and z-coordinates. Hence f has a constrained maximum at $(10, 10, 10)$ which is $f(10, 10, 10) = (10)(10)(10) = 1000$.

17. We must minimize

$$C(x, y) = 50 + 3x^2 + 4y^2$$

subject to the constraint $x + y = 70$. We first rewrite the constraint as

$$g(x, y) = x + y - 70 = 0$$

and then form the function

$$F(x, y, \lambda) = C(x, y) + \lambda g(x, y)$$
$$= 50 + 3x^2 + 4y^2 + \lambda(x + y - 70)$$

Now solve the system of equations

$$F_x(x, y, \lambda) = 6x + \lambda = 0$$
$$F_y(x, y, \lambda) = 8y + \lambda = 0$$
$$F_\lambda(x, y, \lambda) = x + y - 70 = 0$$

The first two equations reduce to

$$x = -\lambda/6$$

and

$$y = -\lambda/8$$

When we substitute these values into the third equation, we obtain

$$-\frac{\lambda}{6} - \frac{\lambda}{8} - 70 = 0$$

Multiplication by -24 yields

$$4\lambda + 3\lambda + 1680 = 0$$
$$7\lambda = -1680$$
$$\lambda = -240$$

When $\lambda = -240$, we have

$$x = 240/6 = 40$$

and

$$y = 240/8 = 30$$

Thus, (40, 30, -240) is the only solution of the system of equations so the constrained minimum value of C(x, y) occurs at (40, 30). This minimum value is

$$C(40,30) = 50 + 3(40)^2 + 4(30)^2 = 8450$$

The minimum weekly manufacturing cost is $8,450.

Exercise Set 8.5, (Page 524)

5. For $f(x, y) = x^2 e^{3y}$ we have

$$\frac{\partial f}{\partial x} = 2xe^{3y}$$

and

$$\frac{\partial f}{\partial y} = 3x^2 e^{3y}$$

Thus the total differential is

$$df = \frac{\partial f}{\partial x} dx + \frac{\partial f}{\partial y} dy$$

$$= (2xe^{3y}) dx + (3x^2 e^{3y}) dy$$

9. For $f(x, y) = \ln(4x^2 y^5) = \ln 4 + 2 \ln x + 5 \ln y$ we have

$$\frac{\partial f}{\partial x} = \frac{2}{x}$$

and

$$\frac{\partial f}{\partial y} = \frac{3}{y}$$

Thus the total differential is

$$df = \frac{\partial f}{\partial x} dx + \frac{\partial f}{\partial y} dy$$

$$= \frac{2}{x} dx + \frac{5}{y} dy$$

13. For $f(x, y) = \sqrt{x^2 + y^3} = (x^2 + y^3)^{1/2}$ we have

$$\frac{\partial f}{\partial x} = \frac{1}{2}(x^2 + y^3)^{-1/2}(2x) = \frac{x}{\sqrt{x^2 + y^3}}$$

and

$$\frac{\partial f}{\partial y} = \frac{1}{2}(x^2 + y^3)^{-1/2}(3y^2) = \frac{3y^2}{2\sqrt{x^2 + y^3}}$$

The total differential is

$$df = \frac{x}{\sqrt{x^2 + y^3}} dx + \frac{3y^2}{2\sqrt{x^2 + y^3}} dy$$

At the point $(-1, 2)$ we have,

$$df = \frac{-1}{\sqrt{(-1)^2 + (2)^3}} dx + \frac{3(2)^2}{2\sqrt{(-1)^2 + (2)^3}} dy$$

$$= \frac{-1}{\sqrt{9}} dx + \frac{12}{2\sqrt{9}} dy$$

$$= \frac{-1}{3} dx + 2\, dy$$

When dx = 0.06 and dy = -0.04, we have

$$df = \frac{-1}{3}(0.06) + 2(-0.04) = -0.02 - 0.08 = -0.10$$

19. The actual change in the function is $\Delta f = f(7.940, 1.030) - f(8, 1)$. To estimate this change we shall use df where

$$df = \frac{\partial f}{\partial x} dx + \frac{\partial f}{\partial y} dy$$

Since $f(x, y) = \sqrt[3]{xy^4} = (xy^4)^{1/3} = x^{1/3} y^{4/3}$ we have,

$$\frac{\partial f}{\partial x} = \frac{1}{3} x^{-2/3} y^{4/3}$$

and

$$\frac{\partial f}{\partial y} = \frac{4}{3} x^{1/3} y^{1/3}$$

At x = 8, y = 1,

$$\frac{\partial f}{\partial x} = \frac{1}{3(\sqrt[3]{8})^2}(1) = \frac{1}{12}$$

and

$$\frac{\partial f}{\partial y} = \frac{4}{3}\sqrt[3]{8}\,(1) = \frac{8}{3}$$

Thus $df = \frac{1}{12} dx + \frac{8}{3} dy$. Taking $\Delta x = -0.06$ and $\Delta y = 0.03$, we have

$$df = \frac{1}{12}(-0.06) + \frac{8}{3}(0.03) = 0.075$$

$$f(7.940, 1.030) = f(8, 1) + \Delta f$$
$$\approx f(8, 1) + df$$
$$f(7.940, 1.030) \approx 2 + 0.075 = 2.075$$

21. To estimate the resulting change in the production we shall use

$$dQ = \frac{\partial Q}{\partial x} dx + \frac{\partial Q}{\partial y} dy$$

Since $Q = 60x^{2/3}y^{1/3}$ we have

$$\frac{\partial Q}{\partial x} = 60y^{1/3}(\tfrac{2}{3}x^{-1/3}) = 40y^{1/3}x^{-1/3}$$

and

$$\frac{\partial Q}{\partial y} = 60x^{2/3}(\tfrac{1}{3}y^{-2/3}) = 20x^{2/3}y^{-2/3}$$

At $x = 8, y = 27,$

$$\frac{\partial Q}{\partial x} = 40\,(27)^{1/3}\,(8)^{-1/3} = \frac{(40)(3)}{2} = 60$$

and

$$\frac{\partial Q}{\partial y} = 20(8)^{2/3}\,(27)^{-2/3} = \frac{(20)(4)}{9} = \frac{80}{9}$$

Thus $dQ = 60\,dx + 80/9\,dy$. Taking $\Delta x = 35/4 - 8 = 3/4$ and $\Delta y = 30 - 27 = 3$, we have

$$dQ = 60\left(\tfrac{3}{4}\right) + \frac{80}{9}(3) = 45 + 26.67 = 71.67$$

The resulting change in production is approximately $71\tfrac{2}{3}$ units of cameras.

25. For $f(x, y, z) = 4xy^3 - 10yz + 8x^2z^3$, we have

$$\frac{\partial f}{\partial x} = 4y^3 + 16xz^3, \qquad \frac{\partial f}{\partial y} = 12xy^2 - 10z,$$

and

$$\frac{\partial f}{\partial z} = -10y + 24x^2 z^2$$

Thus the total differential is

$$df = \frac{\partial f}{\partial x}dx + \frac{\partial f}{\partial y}dy + \frac{\partial f}{\partial z}dz$$

$$= (4y^3 + 16xz^3)dx + (12xy^2 - 10z)dy + (-10y + 24x^2 z^2)dz$$

35. An estimate of the change in the function will be

$$df = \frac{\partial f}{\partial x}dx + \frac{\partial f}{\partial y}dy + \frac{\partial f}{\partial z}dz$$

For $f(x, y, z) = \frac{x\sqrt{y}}{z}$, we have

$$\frac{\partial f}{\partial x} = \frac{\sqrt{y}}{z}, \quad \frac{\partial f}{\partial y} = \frac{x}{2z\sqrt{y}}$$

and

$$\frac{\partial f}{\partial z} = \frac{-x\sqrt{y}}{z^2}$$

At $x = 3$, $y = 4$, and $z = 5$, we have

$$\frac{\partial f}{\partial x} = \frac{\sqrt{4}}{5} = \frac{2}{5}, \qquad \frac{\partial f}{\partial y} = \frac{3}{10\sqrt{4}} = \frac{3}{20},$$

and

$$\frac{\partial f}{\partial z} = \frac{-3\sqrt{4}}{(5)^2} = \frac{-6}{25}$$

Thus

$$df = \frac{2}{5}dx + \frac{3}{20}dy - \frac{6}{25}dz$$

Now $\Delta x = 3.01 - 3 = 0.01$, $\Delta y = 3.97 - 4 = -0.03$, and $\Delta z = 5.02 - 5 = 0.02$. Thus

$$df = \frac{2}{5}(0.01) + \frac{3}{20}(-0.03) - \frac{6}{25}(0.02)$$

= 0.0040 - 0.0045 - 0.0048 = -0.0053

Exercise Set 8.6, (Page 537)

3. Given

$$\int_0^1 \left(\int_0^1 e^{x-y} \, dx\right) dy$$

We first evaluate the inner integral

$$\int_0^1 e^{x-y} \, dx$$

regarding y as a constant. Thus

$$\int_0^1 e^{x-y} \, dy = e^{x-y}\Big]_{x=0}^{x=1} = e^{1-y} - e^{0-y} = e^1 e^{-y} - e^{-y}$$
$$= e^{-y}[e-1]$$

We now evaluate the outer integral

$$\int_0^1 e^{-y}[e-1] \, dy = (e-1) - e^{-y}\Big]_{y=0}^{y=1}$$

$$= (e-1)(-e^{-1} + e^0) = (e-1)\left(1 - \frac{1}{e}\right)$$

$$= e - 1 = 1 + \frac{1}{e} = e - 2 + \frac{1}{e}$$

5. Given

$$\int_1^e \left(\int_1^2 \frac{x}{y} \, dy\right) dx$$

We first evaluate the inner integral

$$\int_1^2 \frac{x}{y} \, dy$$

regarding x as a constant. Thus

$$\int_1^2 \frac{x}{y}\, dy = x \ln y\Big]_{y=1}^{y=2} = x(\ln 2 - \ln 1) = x \ln 2$$

We now evaluate the outer integral

$$\int_1^e x \ln 2\, dx = \ln 2 \int_1^3 x\, dx = \ln 2 [\frac{x^2}{2}]\Big]_{x=1}^{x=e}$$

$$= \ln 2 [\frac{e^2}{2} - \frac{1}{2}] = \frac{\ln 2}{2}(e^2 - 1)$$

13. The double integral of $f(x, 4) = x + 2y$ over the region R, where R is the rectangle defined by $0 \le x \le 2$, $1 \le y \le 3$ can be evaluated by the iterated integral

$$\int_1^3 (\int_0^2 (x + 2y)\, dx)\, dy$$

We first evaluate the inner integral

$$\int_0^2 (x + 2y)\, dx$$

regarding y as a constant. Thus

$$\int_0^2 (x + 2y)\, dx = \frac{x^2}{2} + 2yx\Big]_{x=0}^{x=2}$$

$$= \frac{(2)^2}{2} + 2y(2) - 0 = 2 + 4y$$

We now evaluate the outer integral

$$\int_1^3 (2 + 4y)\, dy = 2y + 2y^2\Big]_{y=1}^{y=3}$$

$$= [2(3) + 2(3)^2] - [2(1) + 2(1)^2]$$

$$= (6 + 18) - (2 + 2) = 20$$

17. The double integral of $f(x, y) = x$ over the region R, where R is the triangle bounded by the lines $y = 2x$, $y = -x$, and $x = 1$ can be evaluated by the iterated integral

$$\int_0^1 \left(\int_{-x}^{2x} x \, dy \right) dx$$

Since the region R is of Type I.

We first evaluate the inner integral

$$\int_{-x}^{2x} x \, dy$$

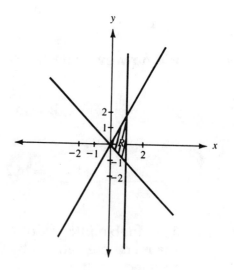

regarding x as a constant. Thus

$$\int_{-x}^{2x} x \, dy = xy \Big]_{y=-x}^{y=2x} = 2x^2 + x^2 = 3x^2$$

We now evaluate the outer integral

$$\int_0^1 3x^2 \, dx = x^3 \Big]_{x=0}^{x=1} = 1$$

21. The volume of the solid lying under the surface $z = 2x + 5y$ and above the region R, where R is the rectangle defined by $2 \leq x \leq 4$, $1 \leq y \leq 3$ can be evaluated by the iterated integral

$$\int_1^3 \left(\int_2^4 (2x + 5y) \, dx \right) dy$$

We first evaluate the inner integral

$$\int_2^4 (2x + 5y) \, dx$$

regarding y as a constant. Thus

$$\int_2^4 (2x + 5y) \, dx = x^2 + 5yx \Big]_{x=2}^{x=4}$$

$$= ((4)^2 + 20y) - ((2)^2 + 10y)$$

$$= (16 + 20y) - (4 + 10y)$$

$$= 12 + 10y$$

We now evaluate the outer integral

$$\int_1^3 (12+10y)\,dy = 12y+5y^2\Big]_{y=1}^{y=3}$$

$$= [12(3)+5(3)^2] - [12(1)+5(1)^2]$$

$$= (36+45) - (12+5) = 81 - 17 = 64$$

27. The volume of the rectangular parallelepiped whose height is 4 and whose base is the rectangle bounded by x = 1, x = 5, y = 2, and y = 4 is evaluated by the iterated integral

$$\int_2^4 \left(\int_1^5 4\,dx\right) dy$$

We first evaluate the inner integral regarding y as a constant. Thus

$$\int_1^5 4\,dx = 4x\Big]_{x=1}^{x=5}$$

$$= 4(5) - 4(1) = 20 - 4 = 16$$

We now evaluate the outer integral

$$\int_2^4 16\,dy = 16y\Big]_{y=2}^{y=4}$$

$$= 16(4) - 16(2) = 64 - 32 = 32$$

We note that the formula for the volume of a rectangular parallelepiped is

$$V = l \cdot w \cdot h$$

In this case l = 5 - 1 = 4, w = 4 - 2 = 2 and h = 4. Thus

$$V = (4)(2)(4) = 32$$

and the answer we obtain using the calculus agrees with the answer from the geometric formula.

31. To find the average value of $f(x, y) = e^x$ over the region R, where R is the triangle in the first quadrant bounded by the line y = 1 - x, the x-axis, and the y-axis, we first must find the area of R.

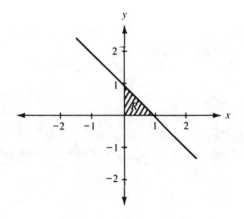

The area of R is

$$\int_0^1 (1-x)\,dx = x - \frac{x^2}{2}\bigg]_0^1$$

$$= 1 - \frac{1}{2} - 0 = \frac{1}{2}$$

Now,

$$f_{ave} = \frac{1}{\text{area of R}} \iint_R f(x,y)\,dx\,dy$$

and

$$\iint_R f(x,y)\,dx\,dy = \int_0^1 \left(\int_0^{1-x} e^x\,dy\right) dx$$

We first evaluate the inner integral

$$\int_0^{1-x} e^x\,dy$$

regarding x as a constant. Thus

$$\int_0^{1-x} e^x\,dy = e^x y\bigg]_{y=0}^{y=1-x} = (1-x)e^x$$

We now evaluate the outer integral.

$$\int_0^1 (1-x)e^x\,dx = \int_0^1 e^x\,dx - \int_0^1 xe^x\,dx$$

We use Formula 40, or integration by parts to evaluate the second integral. We have

$$\int_0^1 e^x \, dx - \int_0^1 xe^x \, dx = e^x \Big]_0^1 - (xe^x - e^x)\Big]_0^1$$

$$= (e - 1) - (e - e + 1) = e - 2$$

Thus the average value is

$$f_{ave} = \frac{1}{\text{area of R}} \iint_R f(x, y) \, dx \, dy$$

$$= \frac{1}{\left(\frac{1}{2}\right)} (e - 2) = 2e - 4 \approx 1.437.$$

Review Exercises, (Page 540)

3. Since $S(P, r, t) = P(1 + rt)$

   (a) $S(4000, 0.08, 3) = 4000(1 + (0.08)(3)) = 4000(1.24) = 4960$

   and

   (b) $S(5000, 0.10, 5) = 5000(1 + (0.10)(5)) = 5000(1.50) = 7500$

7. Since $f(x, y) = x^2 e^{x-y^2}$

$$f_x = x^2 \frac{\partial}{\partial x}[e^{x-y^2}] + e^{x-y^2} \frac{\partial}{\partial x}[x^2]$$

$$= x^2 e^{x-y^2} \frac{\partial}{\partial x}[x - y^2] + e^{x-y^2}[2x]$$

$$= x^2 e^{x-y^2} + 2xe^{x-y^2} = [x^2 + 2x]e^{x-y^2}$$

$$f_y = x^2 \frac{\partial}{\partial y}[e^{x-y^2}] + e^{x-y^2} \frac{\partial}{\partial y}[x^2]$$

$$= x^2 e^{x-y^2} \frac{\partial}{\partial y}[x - y^2] + e^{x-y^2}(0)$$

240   Study Guide

$$= x^2 e^{x-y^2}(-2y) = -2yx^2 e^{x-y^2}$$

and

$$f_{xy} = e^{x-y^2} \frac{\partial}{\partial y}[x^2+2x] + [x^2+2x]\frac{\partial}{\partial y}[e^{x-y^2}]$$

$$= e^{x-y^2}(0) + [x^2+2x] e^{x-y^2} \frac{\partial}{\partial y}[x-y^2]$$

$$= [x^2+2x] e^{x-y^2}(-2y) = -2y[x^2+2x] e^{x-y^2}$$

13. To find the critical points, we determine f_x and f_y and set them equal to zero. In this case

$$f_x = 2x + 3y + 1$$
$$f_y = 3x + 12y - 3$$

Thus we must solve the system of equations

$$2x + 3y + 1 = 0$$
$$3x + 12y - 3 = 0$$

Multiplying the top equation by -4 and adding yields

$$\begin{array}{r} -8x - 12y - 4 = 0 \\ \underline{3x + 12y - 3 = 0} \\ -5x \quad\quad - 7 = 0 \end{array}$$

or

$$x = -\frac{7}{5}$$

Substituting this in the bottom equation yields

$$3(-\frac{7}{5}) + 12y - 3 = 0$$

$$12y = 3 + \frac{21}{5} = \frac{15}{5} + \frac{21}{5} = \frac{36}{5}$$

or

$$y = \frac{36}{5}(\frac{1}{12}) = \frac{3}{5}$$

Thus $(-\frac{7}{5}, \frac{3}{5})$ is the critical point of f.

15. Since $f(x, y) = 3x^2 + 2xy - y^2$, we have

    $f_x = 6x + 2y$
    $f_y = 2x - 2y$
    $f_{xx} = 6$
    $f_{yy} = -2$
    $f_{xy} = 2$

    To find the critical points, we must solve the system of equations

    $6x + 2y = 0$
    $2x - 2y = 0$

    Adding yields $8x = 0$ or $x = 0$. Substituting this value into the first equation yields $6(0) + 2y = 0$ or $y = 0$. Thus $(0, 0)$ is the only critical point of f.

    Since $f_{xx}(0, 0) = 6$, $f_{yy}(0, 0) = -2$ and $f_{xy}(0, 0) = 2$, we have

    $M = f_{xx}(0, 0) f_{yy}(0, 0) - [f_{xy}(0, 0)]^2$

    $= (6)(-2) - (2)^2 = -16$

    Since $M < 0$, we conclude from the second partials test that f has no extremum at $(0, 0)$. Thus f has no relative extrema.

21. We first rewrite the contraint as $g(x, y) = x + y + z - 8 = 0$ and form the new function

    $F(x, y, z, \lambda) = x^2 + y^2 + z^2 + \lambda(x + y + z - 8)$

    We next solve the system

    $F_x (x, y, z, \lambda) = 2x + \lambda = 0$
    $F_y (x, y, z, \lambda) = 2y + \lambda = 0$
    $F_z (x, y, z, \lambda) = 2z + \lambda = 0$
    $F_\lambda (x, y, z, \lambda) = x + y + z - 8 = 0$

    From the first three equations we see that $x = y = z = -\lambda/2$. Substituting into the fourth equation gives

$$-\frac{\lambda}{2}-\frac{\lambda}{2}-\frac{\lambda}{2}-8=0$$

or

$$-\lambda-\lambda-\lambda-16=0$$
$$-3\lambda=16$$
$$\lambda=-\frac{16}{3}$$

so that

$$x=y=z=\frac{16}{6}=\frac{8}{3}$$

Thus $(\frac{8}{3},\frac{8}{3},\frac{8}{3},-\frac{16}{3})$ is the solution of the system, and consequently the constrained minimum value occurs at $(\frac{8}{3},\frac{8}{3},\frac{8}{3})$. Since $f(\frac{8}{3},\frac{8}{3},\frac{8}{3})=(\frac{8}{3})^2+(\frac{8}{3})^2+(\frac{8}{3})^2=3(\frac{64}{9})=\frac{64}{3}$, the constrained minium value is $\frac{64}{3}$.

25. For $f(x, y, z) = x^2y^5 - 4yz^3$ we have

$$\frac{\partial f}{\partial x}=2xy^5 \qquad \frac{\partial f}{\partial y}=5x^2y^4-4z^3$$

and

$$\frac{\partial f}{\partial z}=-12yz^2$$

The total differential is

$$df=\frac{\partial f}{\partial x}dx+\frac{\partial f}{\partial y}dy+\frac{\partial f}{\partial z}dz$$

$$=(2xy^5)\,dx+(5x^2y^4-4z^3)\,dy-12y\,z^2\,dz$$

At the point (3, -1, 2) we have

$$df=2(3)(-1)^5\,dx+[5(3)^2(-1)^4-4(2)^3]\,dy-12(-1)(2)^2\,dz$$

$$df = -6 \, dx + (45 - 32) \, dy + 48 \, dz$$
$$df = -6 \, dx + 13 \, dy + 48 \, dz$$

When $dx = 0.03$, $dy = 0.02$, and $dz = -0.10$ we have

$$df = (-6)(0.03) + 13(0.02) + 48(-0.10)$$
$$= -0.18 + 0.26 - 4.80 = -4.72$$

27. Since $R(x, y) = 2 + 3x^2 + 2xy$,

    (a) the instantaneous rate of change of R with respect to x is

    $$\frac{\partial R}{\partial x} = 6x + 2y$$

    (b) the instantaneous rate of change of R with respect to y is

    $$\frac{\partial R}{\partial y} = 2x$$

33. We must maximize the profit

    $$P(x, y, z) = xz + yz + xy$$

    subject to the expenditure constraint

    $$x + y + z = 60{,}000$$

    We first rewrite the constraint as

    $$g(x, y, z) = x + y + z - 60{,}000 = 0$$

    and then form the function

    $$F(x, y, z, \lambda) = P(x, y, z) + \lambda g(x, y, z)$$

    $$= xz + yz + xy + \lambda(x + y + z - 60{,}000)$$

    Next, we solve the system

    $$F_x(x, y, z, \lambda) = z + y + \lambda = 0$$
    $$F_y(x, y, z, \lambda) = z + x + \lambda = 0$$
    $$F_z(x, y, z, \lambda) = x + y + \lambda = 0$$
    $$F_\lambda(x, y, z, \lambda) = x + y + z - 60{,}000 = 0$$

    Subtracting the second equation from the first equation yields

$$y - x = 0 \quad \text{or} \quad y = x$$

Subtracting the third equation from the first equation yields

$$z - x = 0 \quad \text{or} \quad z = x$$

Substituting $z = y = x$ into the fourth equation yields

$$x + x + x - 60{,}000 = 0$$

or

$$x = 20{,}000$$

When $x = 20{,}000$, we have

$$y = 20{,}000,$$
$$z = 20{,}000,$$

and

$$\lambda = -40{,}000$$

Now $(20{,}000, 20{,}000, 20{,}000, -40{,}000)$ is the only solution of the system of equations. The profit is maximized subject to the expenditure constraint when \$20,000 are spent on each resource.

41. Given

$$\int_1^3 \left(\int_0^x e^{y/x} \, dy \right) dx$$

We first evaluate the inner integral

$$\int_0^x e^{y/x} \, dy$$

regarding x as a constant. Thus

$$\int_0^x e^{y/x} \, dy = x e^{y/x} \Big]_{y=0}^{y=x} = xe - x(e-1)$$

We now evaluate the outer integral.

$$\int_1^3 x(e-1) \, dx = (e-1) \frac{x^2}{2} \Big]_{x=1}^{x=3} = (e-1)\left(\frac{9}{2} - \frac{1}{2}\right) = 4(e-1)$$

45. The volume of the solid below the surface $z = xe^y$ and above the region R, where R is the region bounded by the lines $y = x$, $y = 1$, $y = 3$, and the y-axis can be evaluated by the iterated integral

$$\int_1^3 \left(\int_0^y xe^y \, dx \right) dy$$

since the region R is of Type II.

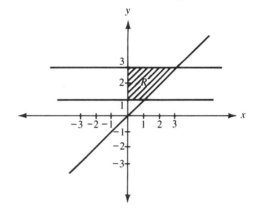

We first evaluate the inner integral

$$\int_0^y xe^y \, dx$$

regarding y as a constant. Thus

$$\int_0^y xe^y \, dx = (e^y)\frac{x^2}{2}\Big]_{x=0}^{x=y} = \frac{y^2}{2}e^y$$

We now evaluate the outer integral

$$\int_1^3 \frac{y^2}{2} e^y \, dy$$

using Formula 42. We have

$$\frac{1}{2}\int_1^3 \frac{y^2}{2} e^y \, dy = \frac{1}{2}\left[y^2 e^y - 2(ye^y - e^y)\right]_{y=1}^{y=3}$$

$$= \frac{e^y}{2}\left[y^2 - 2y + 2\right]_{y=1}^{y=3}$$

$$= \frac{e^3}{2}[9 - 6 + 2] - \frac{e}{2}[1 - 2 + 2]$$

$$= \frac{5e^3}{2} - \frac{e}{2} = \frac{e}{2}[5e^2 - 1]$$

Chapter Test, (Page 543)

1. Since $f(x, y) = \dfrac{x + 2y}{x - y}$

    (a) $f(-1, 2) = \dfrac{-1 + 2(2)}{-1 - 2} = \dfrac{-1 + 4}{-3} = \dfrac{3}{-3} = -1$

    (b) $f(0, 1) = \dfrac{0 + 2(-1)}{0 - (-1)} = \dfrac{-2}{1} = -2$

    (c) $f(3, -5) = \dfrac{3 + 2(-5)}{3 - (-5)} = \dfrac{3 - 10}{8} = \dfrac{-7}{8}$

    (d) The function is defined for all values (x, y) at which the denominator is not equal to zero. Hence the domain consists of all points (x, y) where $x \neq y$.

3. Since $f(x, y) = ye^{-2x} + 4x^2$

$$f_x = y \dfrac{\partial}{\partial x}[e^{-2x}] + e^{-2x}\dfrac{\partial}{\partial x}[y] + \dfrac{\partial}{\partial x}[4x^2]$$

$$= ye^{-2x}\dfrac{\partial}{\partial x}[-2x] + e^{-2x}(0) + 8x$$

$$= ye^{-2x}(-2) + 8x$$

$$= -2ye^{-2x} + 8x$$

$$f_y = y\dfrac{\partial}{\partial y}[e^{-2x}] + e^{-2x}\dfrac{\partial}{\partial y}[y] + \dfrac{\partial}{\partial y}[4x^2]$$

$$= y(0) + e^{-2x} + 0$$

$$= e^{-2x}$$

$$f_{xx} = -2y\dfrac{\partial}{\partial x}[e^{-2x}] + e^{-2x}\dfrac{\partial}{\partial x}[-2y] + \dfrac{\partial}{\partial x}[8x]$$

$$= -2ye^{-2x}\dfrac{\partial}{\partial x}(-2x) + e^{-2x}(0) + 8$$

$$= -2ye^{-2x}(-2) + 8$$

$$= 4ye^{-2x} + 8$$

$$f_{xy} = -2y\frac{\partial}{\partial y}[e^{-2x}] + e^{-2x}\frac{\partial}{\partial y}[-2y] + \frac{\partial}{\partial y}[8x]$$

$$= -2y(0) + e^{-2x}(-2) + 0$$

$$= -2e^{-2x}$$

$$f_{yx} = \frac{\partial}{\partial x}[e^{-2x}] = e^{-2x}\frac{\partial}{\partial x}[-2x] = -2e^{-2x}$$

$$f_{yy} = \frac{\partial}{\partial y}[e^{-2x}] = 0$$

5. For $f(x, y) = x^3(3y^2 - x)$ we have

$$\frac{\partial f}{\partial x} = x^3\frac{\partial}{\partial x}[3y^2 - x] + (3y^2 - x)\frac{\partial}{\partial x}[x^3]$$

$$= x^3(-1) + (3y^2 - x)(3x^2) = -x^3 + 9x^2y^2 - 3x^3$$

$$= -4x^3 + 9x^2y^2$$

$$\frac{\partial f}{\partial y} = x^3\frac{d}{dy}[3y^2 - x] + (3y^2 - x)\frac{\partial}{\partial y}[x^3]$$

$$= x^3(6y) + (3y^2 - x)(0) = 6x^3y$$

The total differential is

$$df = \frac{\partial f}{\partial x}dx + \frac{\partial f}{\partial y}dy$$

$$= (-4x^3 + 9x^2y^2)\,dx + 6x^3y\,dy$$

At the point (1, -2) we have

$$df = [-4(1)^3 + 9(1)^2(-2)^2]\,dx + 6(1)^3(-2)\,dy$$

$$= (-4 + 36)\,dx - 12\,dy$$

$$df = 32\,dx - 12\,dy$$

When $dx = -0.01$ and $dy = 0.02$ we have

$$df = 32(-0.01) - 12(0.02)$$

$$= -0.32 - 0.24 = -0.56$$

7. We first rewrite the constraint as

$$g(x, y) = x^2 + 3y^2 - 3 = 0$$

Next, form the new function

$$f(x, y, \lambda) = x - y + \lambda(x^2 + 3y^2 - 3)$$

Now, solve the system

$$F_x(x, y, \lambda) = 1 + 2\lambda x = 0$$
$$F_y(x, y, \lambda) = -1 + 6\lambda y = 0$$
$$F_\lambda(x, y, \lambda) = x^2 + 3y^2 - 3 = 0$$

The first two equations reduce to

$$x = -1/2\lambda$$
$$y = 1/6\lambda$$

Substitution into the third equation yields

$$(-\frac{1}{2\lambda})^2 + 3(\frac{1}{6\lambda})^2 - 3 = 0$$

$$\frac{1}{4\lambda^2} + \frac{3}{36\lambda^2} - 3 = 0$$

Multiplying by $36\lambda^2$ gives

$$9 + 3 - 108\lambda^2 = 0$$
$$12 = 108\lambda^2$$
$$\frac{12}{108} = \lambda^2$$
$$\frac{1}{9} = \lambda^2$$
$$\lambda = \pm\frac{1}{3}$$

When $\lambda = \frac{1}{3}$, $x = -\frac{1}{2(1/3)} = -\frac{3}{2}$, $y = \frac{1}{6(1/3)} = \frac{1}{2}$

When $\lambda = -\frac{1}{3}$, $x = -\frac{1}{2(-1/3)} = \frac{3}{2}$, $y = \frac{1}{6(-1/3)} = -\frac{1}{2}$

Thus $(-\frac{3}{2}, \frac{1}{2}, \frac{1}{3})$ and $(\frac{3}{2}, -\frac{1}{2}, -\frac{1}{3})$ are solutions of the system of equations. Now since

$$f(x, y) = x - y$$

$$f(-\frac{3}{2}, \frac{1}{2}) = -\frac{3}{2} - \frac{1}{2} = -\frac{4}{2} = -2$$

and

$$f(\frac{3}{2}, -\frac{1}{2}) = \frac{3}{2} - (-\frac{1}{2}) = \frac{4}{2} = 2$$

Hence the constrained maximum value of 2 occurs at $(\frac{3}{2}, -\frac{1}{2})$ and the constrained minimum value of -2 occurs at $(-\frac{3}{2}, \frac{1}{2})$.

9. We wish to maximize the profit $P(x, y) = 4xy + 20x + 16y - 3x^2 - 2y^2$ over the region $x \geq 0$, $y \geq 0$. Now

$$P_x(x, y) = 4y + 20 - 6x$$

and

$$P_y(x, y) = 4x + 16 - 4y$$
$$P_{xx}(x, y) = -6$$
$$P_{yy}(x, y) = -4$$
$$P_{xy}(x, y) = 4$$

To find the critical point, we must solve the system

$$4y + 20 - 6x = 0$$
$$4x + 16 - 4y = 0$$

Rewriting

$$-6x + 4y = -20$$
$$4x - 4y = -16$$

Adding the equations yields

$$-2x = -36$$

or

$$x = 18$$

Substitution into the top equation gives

$$-6(18) + 4y = -20$$
$$-108 + 4y = -20$$
$$4y = 88$$
$$y = 22$$

Thus (18, 22) is the only critical point. Now $P_{xx}(18, 22) = -6$, $P_{yy}(18, 22) = -4$, and $P_{xy}(18, 22) = 4$, so that

$$M = P_{xx}(18, 22)\, P_{yy}(18, 22) - [P_{xy}(18, 22)]^2$$

$$= (-6)(-4) - (4)^2 = 24 - 16 = 8$$

Since $M > 0$ and $P_{xx}(18, 22) < 0$, the profit is a maximum at (18, 22). Thus plant A should produce 18 hundred coats weekly, and plant B should produce 22 hundred coats weekly.

11. The volume is given by

$$V = xyz$$

We want to find ΔV when $x = 12$ feet, $y = 8$ feet, $z = 4$ feet, and $\Delta x = \Delta y = \Delta z = 6$ inches $= 0.5$ feet. We will use dV to approximate ΔV. Now

$$dV = \frac{\partial V}{\partial x} dx + \frac{\partial V}{\partial y} dy + \frac{\partial V}{\partial z} dz$$

$$= yz\, dx + xz\, dy + xy\, dz$$

$$dV = (8)(4)(0.5) + (12)(4)(0.5) + (12)(8)(0.5)$$

$$dV = 16 + 24 + 48 = 88$$

Thus the vault contains approximately 88 cubic feet of concrete.

13. The volume of the solid below the surface z = 3x + y and over the rectangular region R with vertices (0, 0), (3, 0), (0, 4) and (3, 4) can be evaluated by the iterated integral

$$\int_0^4 \left(\int_0^3 (3x+y)\,dx\right) dy$$

We first evaluate the inner integral

$$\int_0^3 (3x+y)\,dx$$

regarding y as a constant. Thus

$$\int_0^3 (3x+y)\,dx = (3x^2 + yx)\Big]_{x=0}^{x=3} = \frac{27}{2} + 3y$$

We now evaluate the outer integral

$$\int_0^4 (\frac{27}{2} + 3y)\,dy = (\frac{27}{2}y + \frac{3y^2}{2})\Big]_{y=0}^{y=4} = \frac{27}{2}(4) + \frac{3(4)^2}{2} - 0 = 54 + 24 = 78$$

9 The Trigonometric Functions (Optional)

Key Ideas for Review

* Angles are measured in degrees and in radians.

* 2π radians $= 360^0$

* $\dfrac{\text{radian measure of } \theta}{\pi \text{ radians}} = \dfrac{\text{degree measure of } \theta}{180^0}$

* The isosceles right triangle:

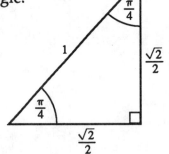

* The $30^0, 60^0, 90^0$, right triangle

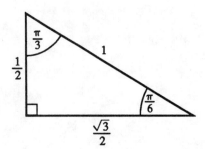

* Let θ be a fixed angle that is in standard position. Let $P(x, y)$ be the point at which the terminal side of θ intersects the unit circle centered at the origin. Then

$$\sin \theta = y, \quad \cos \theta = x$$

$$\tan \theta = \frac{y}{x} \qquad\qquad x \neq 0$$

$$\csc \theta = \frac{1}{y} \qquad\qquad y \neq 0$$

$$\sec \theta = \frac{1}{x} \qquad\qquad x \neq 0$$

$$\cot \theta = \frac{x}{y} = \frac{1}{\tan \theta}, \qquad\qquad y \neq 0$$

* $\tan \theta = \frac{\sin \theta}{\cos \theta}; \sec \theta = \frac{1}{\cos \theta}; \csc \theta = \frac{1}{\sin \theta}; \cot \theta = \frac{1}{\tan \theta}$

* $\sin(\theta + 2\pi) = \sin \theta; \cos(\theta + 2\pi) = \cos \theta; \tan(\theta + \pi) = \tan \theta$

* $\sin(-\theta) = -\sin \theta; \cos(-\theta) = \cos \theta; \tan(-\theta) = -\tan \theta$

* $\sin^2 \theta + \cos^2 \theta = 1$

* If θ is an acute angle in a right traingle, then

* $\sin \theta = \frac{\text{opposite}}{\text{hypotenuse}}$

* $\cos \theta = \frac{\text{adjacent}}{\text{hypotenuse}}$

* $\tan \theta = \frac{\text{opposite}}{\text{adjacent}}$

* If y = sin u and u is a differentiable function of x, then

$$\frac{dy}{dx} = \frac{d}{dx}[\sin u] = \cos u \cdot \frac{du}{dx}$$

* If y = cos u and u is a differentiable function of x, then

$$\frac{dy}{dx} = \frac{d}{dx}[\cos u] = -\sin u \cdot \frac{du}{dx}$$

* If y = tan u and u is a differentiable function of x, then

$$\frac{dy}{dx} = \frac{d}{dx}[\tan u] = \sec^2 u \cdot \frac{du}{dx}$$

* $\int \sin u \, du = -\cos u + C$

* $\int \cos u \, du = \sin u + C$

* $\int \tan u \, du = -\ln |\cos u| + C$

Exercise Set 9.1, (Page 560)

7. Using the conversion formula, we have

$$\frac{80^0}{180^0} = \frac{\theta}{\pi}$$

or

$$\theta = \frac{80}{180} \pi = \frac{4}{9} \pi$$

9. Using the conversion formula, we have

$$\frac{-150^0}{180^0} = \frac{\theta}{\pi}$$

or

$$\theta = \frac{-150}{180} \pi = \frac{-5}{6} \pi$$

13. Using the conversion formula, we have

$$\frac{\frac{4}{3}\pi}{\pi} = \frac{\theta}{180^0}$$

or

$$\theta = \frac{4}{3}(180^0) = 240^0$$

15. Using the conversion formula, we have

$$\frac{\frac{-7\pi}{12}}{\pi} = \frac{\theta}{180^0}$$

or

$$\theta = \frac{-7}{12}(180^0) = -105^0$$

21. Now $\sin(-\pi) = -\sin \pi = (-1)(0) = 0$

23. Now

$$\tan\left(\frac{-\pi}{4}\right) = \frac{\sin\left(\frac{-\pi}{4}\right)}{\cos\left(\frac{-\pi}{4}\right)} = \frac{-\sin\frac{\pi}{4}}{\cos\frac{\pi}{4}} = \frac{-\frac{\sqrt{2}}{2}}{\frac{\sqrt{2}}{2}} = -1$$

31. $\sin \theta = \dfrac{\text{length of side opposite } \theta}{\text{length of hypotenuse}} = \dfrac{6}{10} = \dfrac{3}{5}$

$\cos \theta = \dfrac{\text{length of side adjacent to } \theta}{\text{length of hypotenuse}} = \dfrac{8}{10} = \dfrac{4}{5}$

$\tan \theta = \dfrac{\sin \theta}{\cos \theta} = \dfrac{3/5}{4/5} = \dfrac{3}{4}$

33. $\sin \theta = \dfrac{\text{length of side opposite } \theta}{\text{length of hypotenuse}} = \dfrac{x}{\sqrt{x^2+1}}$

$\cos \theta = \dfrac{\text{length of side adjacent to } \theta}{\text{length of hypotenuse}} = \dfrac{1}{\sqrt{x^2+1}}$

$\tan \theta = \dfrac{\sin \theta}{\cos \theta} = \dfrac{x/\sqrt{x^2+1}}{1/\sqrt{x^2+1}} = x$

45. Let h denote the height of the kite in meters. From the figure below, we have

$$\frac{h}{150} = \sin 35^0$$

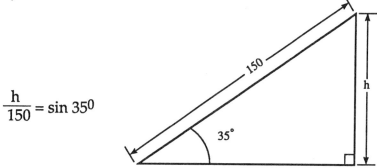

so that h = 150 sin 35⁰ = 86 meters.

47. Let x denote the distance (in feet) from the base of the tower to the fire

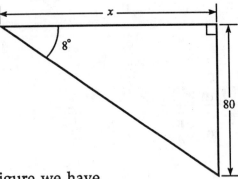

then from the figure we have

$$\tan 8^0 = \frac{\text{length of side opposite } 8^0}{\text{length of side adjacent to } 8^0} = \frac{80}{x}$$

Thus

$$x = \frac{80}{\tan 8^0} = 569.2 \text{ feet}$$

Exercise Set 9.2, (Page 569)

1. $\dfrac{dy}{dx} = \dfrac{d}{dx}[\cos 3x] = -\sin 3x \dfrac{d}{dx}[3x] = -3 \sin 3x$

5. Using the power rule, we have

$$\frac{dy}{dx} = \frac{d}{dx}[(\sin x)^3] = 3(\sin x)^2 \frac{d}{dx}[\sin x]$$
$$= 3(\sin x)^2 \cos x$$
$$= 3 \sin^2 x \cos x$$

15. Using the product rule, we have

$$\frac{dy}{dx} = \frac{d}{dx}[x^3 \tan x] = x^3 \frac{d}{dx}[\tan x] + \tan x \frac{d}{dx}[x^3]$$
$$= x^3 \sec^2 x + (\tan x)(3x^2)$$
$$= x^3 \sec^2 x + 3x^2 \tan x$$

9: The Trigonometric Functions.. 257

19. $\dfrac{dy}{dx} = \dfrac{d}{dx}[\cos(e^{2x})] = -\sin(e^{2x}) \dfrac{d}{dx}[e^{2x}]$

   $= -\sin(e^{2x}) \, 2e^{2x} = -2e^{2x} \sin(e^{2x})$

21. Using the product rule we have

   $\dfrac{dy}{dx} = \dfrac{d}{dx}[e^x \sin x] = e^x \dfrac{d}{dx}[\sin x] + \sin x \dfrac{d}{dx}[e^x]$

   $= e^x \cos x + (\sin x)e^x = e^x(\sin x + \cos x)$

25. Using the product and power rules, we have

   $\dfrac{dy}{dx} = \dfrac{d}{dx}[\sin^2 x \cos x]$

   $= \sin^2 x \dfrac{d}{dx}[\cos x] + \cos x \dfrac{d}{dx}[\sin^2 x]$

   $= \sin^2 x \, (-\sin x) + \cos x \, 2\sin x \dfrac{d}{dx}[\sin x]$

   $= -\sin^3 x + \cos x(2 \sin x \, \cos x)$

   $= 2 \cos^2 x \, \sin x - \sin^3 x$

31. Now

   $\dfrac{d}{dx}[\csc x] = \dfrac{d}{dx}[\dfrac{1}{\sin x}]$

   $= \dfrac{d}{dx}[(\sin x)^{-1}]$

   $= (-1)(\sin x)^{-2} \dfrac{d}{dx}[\sin x]$

   $= -\dfrac{1}{(\sin x)^2} \cos x$

   $= -\dfrac{1}{\sin x} \cdot \dfrac{\cos x}{\sin x}$

   $= -\csc x \cot x$

258 Study Guide

33. By the quotient rule, we have

$$\frac{d}{dx}[\cot x] = \frac{d}{dx}[\frac{\cos x}{\sin x}]$$

$$= \frac{\sin x \frac{d}{dx}[\cos x] - \cos x \frac{d}{dx}[\sin x]}{(\sin x)^2}$$

$$= \frac{\sin x (-\sin x) - \cos x (\cos x)}{(\sin x)^2}$$

$$= -\frac{\sin^2 x + \cos^2 x}{\sin^2 x}$$

but for any value of x,

$$\sin^2 x + \cos^2 x = 1$$

so that

$$\frac{d}{dx}[\cot x] = -\frac{1}{\sin^2 x} = -(\frac{1}{\sin x})^2 = -\csc^2 x$$

41. The slope of the tangent line to the graph of $f(x) = \cos 2x$ at $x = \pi/6$ is $f'(\pi/6)$. Now

$$f'(x) = \frac{d}{dx}[\cos 2x] = -\sin 2x \frac{d}{dx}[2x]$$

$$= -2 \sin 2x$$

so that

$$f'(\pi/6) = -2 \sin (2 \cdot \frac{\pi}{6}) = -2 \sin \frac{\pi}{3} = -2\frac{\sqrt{3}}{2} = -\sqrt{3}$$

45. The instantaneous rate at which the amount of waste is dumped each day is changing after 6 months is given by A'(6). We have

$$A'(t) = 12 \cos (\frac{\pi}{6} t) \cdot \frac{\pi}{6}$$

$$A'(t) = (12) (\frac{\pi}{6}) \cos (\frac{\pi}{6} t)$$

9: The Trigonometric Functions.. 259

$$A'(t) = 2\pi \cos\left(\frac{\pi}{6} t\right)$$

so that

$$A'(6) = 2\pi \cos\left[\frac{\pi}{6}(6)\right]$$

$$= 2\pi \cos \pi = 2\pi(-1) = -2\pi$$

Thus the rate is decreasing by 2000π gallons per day per month.

Exercise Set 9.3, (Page 576)

3. We use the method of substitution with $u = x + 4$ so that $du = dx$ and

   $$\int \cos(x+4)\, dx = \int \cos u\, du = \sin u + C$$
   $$= \sin(x+4) + C$$

7. We use the method of substitution with $u = \sin x$ so that $du = \cos x\, dx$ and

   $$\int \sin x \cos x\, dx = \int u\, du = \frac{1}{2}u^2 + C$$

   $$= \frac{1}{2}\sin^2 x + C$$

15. We use the method of substitution with $u = \sin x$ so that $du = \cos x\, dx$ and

    $$\int \frac{\cos x}{\sqrt{\sin x}}\, dx = \int \frac{1}{\sqrt{u}}\, du = \int u^{-1/2}\, du$$

    $$= \frac{u^{1/2}}{1/2} + C = 2\sqrt{u} + C$$

    $$= 2\sqrt{\sin x} + C$$

25. We first use the method of substitution with $u = 2x$ and $du = 2\, dx$ to evaluate the indefinite integral. Noting that $dx = du/2$, we have

$$\int \sin 2x\, dx = \int \sin u\, (\tfrac{1}{2})du = \tfrac{1}{2}\int \sin u\, du$$

$$= -\tfrac{1}{2}\cos u + C = -\tfrac{1}{2}\cos 2x + C$$

Thus,

$$\int_{\pi/4}^{3\pi/4} \sin 2x\, dx = -\tfrac{1}{2}\cos 2x \Big]_{\tfrac{\pi}{4}}^{\tfrac{3\pi}{4}}$$

$$= -\tfrac{1}{2}\cos(2\cdot\tfrac{3\pi}{4}) + \tfrac{1}{2}\cos(2\cdot\tfrac{\pi}{4})$$

$$= -\tfrac{1}{2}\cos\tfrac{3\pi}{2} + \tfrac{1}{2}\cos\tfrac{\pi}{2}$$

$$= -\tfrac{1}{2}\cdot 0 + \tfrac{1}{2}\cdot 0$$

$$= 0$$

29. We first determine the indefinite integral using the method of substitution with $u = x^4$ so that $du = 4x^3\, dx$ and $x^3\, dx = du/4$. Now

$$\int x^3 \sin(x^4)\, dx = \int \tfrac{1}{4}\sin u\, du$$

$$= -\tfrac{1}{4}\cos u + C$$

$$= -\tfrac{1}{4}\cos(x^4) + C$$

so that

$$\int_0^{\pi} x^3 \sin(x^4)\, dx = -\tfrac{1}{4}\cos(x^4) \Big]_0^{\pi}$$

$$= -\tfrac{1}{4}\cos(\pi^4) + \tfrac{1}{4}\cos(0^4)$$

$$= \frac{1}{4}(1 - \cos(\pi^4))$$

31. We first determine the indefinite integral using integration by parts. Let

$$u(x) = x \text{ and } v'(x) = \cos 2x$$

so that

$$u'(x) = 1 \text{ and } v(x) = \int v'(x)\,dx = \int \cos 2x\,dx = \frac{1}{2}\sin 2x$$

Then

$$\int x \cos 2x\,dx = \int u(x)\,v'(x)\,dx$$

$$= u(x)\,v(x) - \int v(x)\,u'(x)\,dx$$

$$= \frac{x}{2}\sin 2x - \int \frac{1}{2}\sin 2x\,dx + C$$

$$= \frac{x}{2}\sin 2x + \frac{1}{4}\cos 2x + C$$

so that

$$\int_0^{\pi/4} x \cos 2x\,dx$$

$$= \left(\frac{x}{2}\sin 2x + \frac{1}{4}\cos 2x\right)\Big]_0^{\frac{\pi}{4}}$$

$$= \frac{\pi/4}{2}\sin\left(2\cdot\frac{\pi}{4}\right) + \frac{1}{4}\cos\left(2\cdot\frac{\pi}{4}\right) - \left[\frac{0}{2}\sin 2\cdot 0 + \frac{1}{4}\cos(2\cdot 0)\right]$$

$$= \frac{\pi}{8}\sin\frac{\pi}{2} + \frac{1}{4}\cos\frac{\pi}{2} - \frac{1}{4}\cos 0$$

$$= \frac{\pi}{8}(1) + \frac{1}{4}(0) - \frac{1}{4}(1)$$

$$= \frac{\pi}{8} - \frac{1}{4} = \frac{\pi - 2}{8}$$

33. The desired area is

262 Study Guide

$$\int_0^{\pi/2} \sin \frac{1}{2} x \, dx$$

We first determine the indefinite integral using the method of substitution with $u = 1/2\, x$ so that $du = 1/2\, dx$ and $2\, du = dx$. Now

$$\int \sin \frac{1}{2} x \, dx = \int \sin u \,(2\, du) = 2\int \sin u \, du$$

$$= -2 \cos u + C = -2 \cos \frac{1}{2} x + C$$

so that

$$\int_0^{\pi/2} \sin \frac{1}{2} x \, dx = -2 \cos \frac{1}{2} x \Big]_0^{\frac{\pi}{2}}$$

$$= -2 \left[\cos \frac{1}{2}(\frac{\pi}{2}) - \cos \frac{1}{2}(0)\right]$$

$$= -2 \left[\cos \frac{\pi}{4} - \cos 0\right]$$

$$= -2 \left[\frac{\sqrt{2}}{2} - 1\right] = -\sqrt{2} + 2$$

Review Exercises, (Page 578)

3. Using the conversion formula, we have

$$\frac{315^0}{180^0} = \frac{\theta}{\pi}$$

or

$$\theta = \frac{315}{180} \pi = \frac{7}{4} \pi$$

5. Using the conversion formula, we have

$$\frac{\frac{11\pi}{6}}{\pi} = \frac{\theta}{180^0}$$

or
$$\theta = \frac{11}{6}(180^0) = 330^0$$

9. Now
$$\tan\left(-\frac{2}{3}\pi\right) = \frac{\sin\left(-\frac{2}{3}\pi\right)}{\cos\left(-\frac{2}{3}\pi\right)}$$

$$= \frac{-\sin\frac{2}{3}\pi}{\cos\frac{2}{3}\pi}$$

$$= \frac{-\sqrt{3}/2}{-1/2}$$

$$= \sqrt{3}$$

17. Note that $y = (\cos x)^2 \sin 2x$. Using the product and power rules, we have

$$\frac{dy}{dx} = (\cos x)^2 \frac{d}{dx}[\sin 2x] + \sin 2x \frac{d}{dx}[(\cos x)^2]$$

$$= (\cos x)^2 \cos 2x \frac{d}{dx}[2x] + \sin 2x (2 \cos x) \frac{d}{dx}[\cos x]$$

$$= 2(\cos^2 x) \cos 2x + 2(\sin 2x) \cos x(-\sin x)$$

$$= 2(\cos^2 x) \cos 2x - 2(\sin 2x) \cos x \sin x$$

21. Now $y = \sqrt{\tan x} = (\tan x)^{1/2}$, so that the power rule yields

$$\frac{dy}{dx} = \frac{1}{2}(\tan x)^{-1/2} \frac{d}{dx}[\tan x]$$

$$= \frac{1}{2}(\tan x)^{-1/2} \sec^2 x = \frac{\sec^2 x}{2\sqrt{\tan x}}$$

23. $\frac{\partial z}{\partial x} = 3 \frac{\partial}{\partial x}[\sin 2x] - 2 \frac{\partial}{\partial x}[\cos 3y]$

$$= 3\cos 2x \frac{\partial}{\partial x}[2x] - 2(0)$$

$$= 6\cos 2x$$

and

$$\frac{\partial z}{\partial y} = 3\frac{\partial}{\partial y}[\sin 2x] - 2\frac{\partial}{\partial y}[\cos 3y]$$

$$= 3(0) - 2(-\sin 3y)\frac{\partial}{\partial y}[3y]$$

$$= 2(\sin 3y)(3)$$

$$= 6\sin 3y$$

27. We use the method of substitution with $u = \cos x$. In this case, $du = -\sin x\, dx$ so that $\sin x\, dx = -du$ and

$$\int e^{\cos x} \sin x\, dx = \int e^u (-1)\, du$$

$$= -\int e^u\, du = -e^u + C$$

$$= -e^{\cos x} + C$$

31. We first determine the indefinite integral using integration by parts. Let

$$u(x) = x \text{ and } v'(x) = \sin 2x$$

so that

$$u'(x) = 1 \text{ and } v(x) = \int v'(x)\, dx = \int \sin 2x\, dx = -\frac{1}{2}\cos 2x$$

Then

$$\int x \sin 2x\, dx = \int u(x) v'(x)\, dx$$

$$= u(x) v(x) - \int v(x) u'(x)\, dx$$

$$= -\frac{x}{2}\cos 2x - \int -\frac{1}{2}\cos 2x\, dx$$

9: The Trigonometric Functions.. 265

$$= -\frac{x}{2}\cos 2x + \frac{1}{2}\int \cos 2x \, dx$$

$$= -\frac{x}{2}\cos 2x + \frac{1}{4}\sin 2x + C$$

Thus,

$$\int_{\pi/2}^{3\pi/2} x \sin 2x \, dx = \left(-\frac{x}{2}\cos 2x + \frac{1}{4}\sin 2x\right)\Big]_{\frac{\pi}{2}}^{\frac{3\pi}{2}}$$

$$= [-\frac{3\pi/2}{2}\cos(2 \cdot 3\pi/2) + \frac{1}{4}\sin(2 \cdot 3\pi/2)]$$

$$- [\frac{-\pi/2}{2}\cos(2 \cdot \pi/2) + \frac{1}{4}\sin(2 \cdot \pi/2)]$$

$$= [-\frac{3\pi}{4}\cos 3\pi + \frac{1}{4}\sin 3\pi] - [-\frac{\pi}{4}\cos \pi + \frac{1}{4}\sin \pi]$$

$$= [-\frac{3\pi}{4}(-1) + \frac{1}{4}(0)] - [-\frac{\pi}{4}(-1) + \frac{1}{4}(0)]$$

$$= \frac{3\pi}{4} - \frac{\pi}{4} = \frac{\pi}{2}$$

33. The region bounded by $y = \sin x$ and the x-axis from $x = -\pi$ to $x = \pi$ is shaded in the figure below

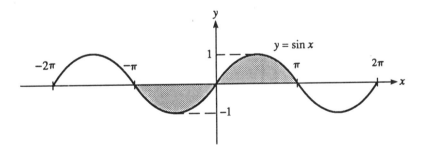

The total area bounded by the curve $y = \sin x$ and the x-axis from $x = -\pi$ to $x = \pi$ is the sum of the area under $[-\pi, 0]$ and over $y = \sin x$ and the area under $y = \sin x$ over $[0, \pi]$. Thus, this total area is

$$-\int_{-\pi}^{0}\sin x \, dx + \int_{0}^{\pi}\sin x \, dx = \cos x\Big]_{-\pi}^{0} + (-\cos x)\Big]_{0}^{\pi}$$

$$= (\cos 0 - \cos(-\pi)) + (-\cos \pi + \cos 0)$$
$$= (1 - (-1)) + (-(-1) + 1)$$
$$= 2 + 2 = 4$$

37. The rate of change of the number of people who have the disease is given (in people per day) by N'(t). Now

$$N'(t) = 100 \frac{d}{dt}[\sin(\frac{\pi t}{10})]$$

$$= 100 \cos(\frac{\pi t}{10}) \frac{d}{dt}[\frac{\pi t}{10}]$$

$$= 100 \frac{\pi}{10} \cos(\frac{\pi t}{10})$$

$$= 10\pi \cos(\frac{\pi t}{10})$$

so that

$$N'(5/2) = 10\pi \cos(\frac{\pi}{10} \cdot \frac{5}{2}) = 10\pi \cos \frac{\pi}{4} = 5\pi\sqrt{2} \approx 22.2$$

and

$$N'(40/3) = 10\pi \cos(\frac{\pi}{10} \cdot \frac{40}{3}) = 10\pi \cos \frac{4\pi}{3} = -5\pi = -15.7$$

(a) After 5/2 days the number of people who have the disease is increasing at a rate of $5\pi\sqrt{2} \approx 22$ people per day.

(b) After 40/3 days the number of people who have the disease is changing at a rate of $-5\pi \approx -16$ people per day.

Chapter Test, (Page 580)

1. (a) Using the conversion formula, we have

$$\frac{150°}{180°} = \frac{\theta}{\pi}$$

or

9: The Trigonometric Functions.. 267

$$\theta = \frac{150}{180}\pi = \frac{5}{6}\pi$$

(b) Using the conversion formula, we have

$$\frac{\frac{7\pi}{3}}{\pi} = \frac{\theta}{180^0}$$

or

$$\theta = \frac{7}{3}(180^0) = 420^0$$

3. Now $y = \cos(3x + \frac{\pi}{2})$, so

$$\frac{dy}{dx} = [-\sin(3x + \frac{\pi}{2})]\frac{d}{dx}(3x + \frac{\pi}{2})$$

$$= [-\sin(3x + \frac{\pi}{2})](3)$$

$$= -3\sin(3x + \frac{\pi}{2})$$

5. Now $y = x^3 \sin x^2$ so that using the product and power rules we have

$$\frac{dy}{dx} = x^3 \frac{d}{dx}[\sin x^2] + (\sin x^2)\frac{d}{dx}[x^3]$$

$$= x^3 \cos x^2 \frac{d}{dx}[x^2] + (\sin x^2)3x^2$$

$$= (x^3 \cos x^2)2x + 3x^2 \sin x^2$$

$$= 2x^4 \cos x^2 + 3x^2 \sin x^2$$

7. We use the method of substitution with $u = \sin x$. Then $du = \cos x\, dx$ so that

$$\int \sqrt{\sin x} \cos x\, dx = \int \sqrt{u}\, du = \int u^{1/2}\, du$$

$$= \frac{2}{3}u^{3/2} + C$$

$$= \frac{2}{3}(\sin x)^{3/2} + C$$

9. We use the method of substitution with $u = 3x - 4$. Then $du = 3\,dx$ or $dx = \frac{1}{3}du$, so that

$$\int \tan(3x-4)\,dx = \int \tan u \left(\frac{1}{3}du\right) = \frac{1}{3}\int \tan u\,du$$
$$= -\frac{1}{3}\ln|\cos u| + C$$
$$= -\frac{1}{3}\ln|\cos(3x-4)| + C$$